Methods in Computational Biology

Methods in Computational Biology

U. S. Raghavender, Ph.D.

Delve
PUBLISHING
www.delvepublishing.com

Methods in Computational Biology

U. S. Raghavender, Ph.D.

Delve Publishing

2010 Winston Park Drive,

2nd Floor

Oakville, ON L6H 5R7

Canada

www.delvepublishing.com

Tel: 001-289-291-7705
001-905-616-2116

Fax: 001-289-291-7601

Email: orders@arclereducation.com

ISBN: 978-1-77361-241-6 (Hardcover)

ABOUT THE AUTHOR

U. S. Raghavender, Ph.D.

Dr. U. S. Raghavender did his Masters in Physics, with specialization in Condensed Matter Physics, followed by his doctoral thesis in X-ray Crystallography of Designed molecules. This was followed by post-doctoral stints in India and abroad, in both academic institutions and non-academic institutions. He has been working on computational approaches of biomolecules, since the past 12 years. He has authored book chapters, published research in internationally acclaimed journals. He also serves as a peer reviewer of research articles published in both national (*Journal of Biosciences*) and international journals (*BMC Bioinformatics*, *Nucleic Acids Research etc.*). He has a long term goal of addressing computational aspects of regulatory mechanisms in cell.

TABLE OF CONTENTS

PREFACE

The field of *computational biology*, or *bioinformatics* as it is often called, was born just a few years ago. It is difficult to pinpoint its exact beginnings, but it is easy to see that the field is currently undergoing rapid, exciting growth. This growth has been fuelled by a revolution in DNA sequencing and mapping technology, which has been accompanied by rapid growth in many related areas of biology and biotechnology. No doubt many exciting breakthroughs are yet to come. All this new DNA and protein sequence data brings with it the tremendously exciting challenge of how to make sense of it: how to turn the raw sequences into information that will lead to new drugs, new advances in health care, and a better overall understanding of how living organisms function. One of the primary tools for making sense of this revolution in sequence data is the **computer**. Computational biology is all about how to use the power of computation to model and understand biological systems and especially biological sequence data.

This book is written keeping in mind the graduates and researchers studying life sciences using computational techniques. Although, it can also be used by undergraduates as a spring board for exploring the field of *Computational Biology*. This could serve as a supplement for people involved in specialized research. With the flood of genomics and proteomics data, it is imperative that students in biology have a basic understanding of the data generated and how it is processed using computation. The algorithms could be graded from simple to complex, based on the assumptions employed. In this book, our focus will be to understand the most commonly used algorithms in biology and also take a peek at the latest developments in genomics based computational advances.

This book is an attempt to bring together in one place some of the latest advances in computational biology. With this in mind, we have included tutorials on many of the key topics in the volume, designed to introduce biological scientists to some of the computational techniques that might otherwise be unfamiliar to them. Many of those tutorials are interweaved with the main body of the chapters, which provides a nice hands-on experience to the reader. It is expected that the reader is familiar with programming languages like **Python** and **R**.

We also want to encourage more computer scientists to get involved in this new field, and with them in mind we included tutorial material on several topics in molecular biology as well. We hope the result is a volume that offers something valuable to a wide range of readers. The only required background is an interest in the exciting and mature field of computational biology. The chapters that follow are broadly grouped into three sections. Loosely speaking, these can be described as an introductory section, a section on DNA sequence analysis, and a section on proteins. The introductory section begins with an overview of some of the main challenges facing computational biology today. This chapter contains a thought-provoking description of problems ranging from gene finding to protein folding, explaining the biological significance and hinting at many of the computational solutions that appear in later chapters. Next is a tutorial on computation, designed primarily for biologists who do not have a formal background in computer science. After reading this chapter, biologists should find many of the later chapters much more accessible. The following chapter provides a tutorial for the other main component of our audience, computational scientists (including computer scientists, mathematicians, physicists, and anyone else who might need some additional biological background) who want to understand the biology that underlies all the research problems described in later chapters.

The second of the three major sections contains work primarily on DNA sequence analysis. Although the techniques covered here are not restricted to DNA sequences, most of the applications described here have been applied to DNA. One chapter is devoted to the eternal problem in computational protein structure prediction from sequences.

The purpose of the present book is to present a representative sample of computational problems in molecular biology and present some efficient algorithms that are offered as solutions to these problems. In fact, the recipes provided here can also be used in day-to-day research endeavours. Every chapter begins with a small introduction followed by explanation of concepts. Computational Code is given in bounded box, and in a font different from the text. References are provided at the end of each chapter, as needed. Each chapter is written to be self-contained. We hope that this will help reader use computers and computational techniques quite effectively in their research work.

Computational molecular biology is expanding fast. Better algorithms are constantly being designed, and new subfields are emerging even as we write this. Within the constraints mentioned above, we did our best to cover what we considered a wide range of topics, and we believe that most of the material presented is of lasting value.

Author

1

COMPUTATIONAL BIOLOGY

Biology is broadly defined as the study of living beings. Biologists study living things by performing experiments, collecting and interpreting data. With the advent of technology in the biological disciplines, we have a lot of experimental data generated within a short time. The sophistication of the experimental procedures and the vast volume of data, requires a concomitant match in exploration using computation. The most commonly cited example for this *dichotomy* is the **data** generated out of genomics experiments.

The notion of "challenge" conjures up images ranging from solving a crystal structure to visualization of the disease progression. The interdisciplinary field of *Bioinformatics* has elements of "challenge" and spans a wide umbrella of interesting topics directly connecting with human health and molecular comprehension of disease fundamentals. In particular, the filed underpins the movement towards the "large scale biology", i.e., ***genomics***.

At present, we can generate a large volume of DNA sequence data in a short space of time. We will be left with a large ocean of adenosines (A), thymines (T), cytosines (C) and guanines (G). *Bioinformatics* is a science to understand biology using computational techniques. Essentially, it is a collection of tools which can be used to explore and answer a few particular questions in biology. It can be put under the umbrella of a much wider field of *Computational Biology*. Computational Biology aims at exploring large amount of biological data with quantitative analytical techniques. This includes developing and implementing efficient algorithms in programming languages directed towards biological questions. We will be focusing our attention on different methods in computational biology in this book.

In a very broad sense ***computational molecular biology*** consists of the development and use of mathematical and computer science techniques to help solve problems in molecular biology. Databases are needed to store all the information that is being generated. Several international sequence databases already exist, but scientists have recognized the need for new database models, given the specific requirements of molecular biology. For example, these databases should be able to record changes in our understanding of molecular sequences as we study them; current models have limitations which cannot satisfactorily address this concern. The understanding of molecular sequences in turn requires new sophisticated techniques of *pattern recognition*, which are being developed by researchers in artificial intelligence. Complex statistical issues have arisen in connection with database searches, and this has surged the creation of new and specific tools.

There is one class of problems, however, for which what is most needed is *efficient algorithms.* An **algorithm**, simply stated, is a

Step-by-step procedure that tries to solve a certain well-defined problem in a limited time bound.

To be *efficient*, an algorithm should not take "too long" to solve a problem, even a large one. The classic example of a problem in molecular biology solvable by an algorithm is sequence comparison:

Given two sequences representing biomolecules, we want to know how similar they are?

This is a problem that must be solved thousands of times every day, so it is desirable that a very efficient algorithm should be employed.

The field of *Bioinformatics* gained attention with the painstaking efforts of a few dedicated scientists who had expertise in the areas of statistics and pattern recognition. Presently, the field welcomes people from varied backgrounds including mathematics, statistics, computer science, linguistics, chemistry, and biology. Naturally, each and every tool has a pitfall(s) in bioinformatics. Users need to have a good layman understanding of what each tool and algorithm does and in what contexts it can be used to answer specific biological questions. There is no need to overemphasize the importance of bioinformatics and computational science in the era of *-omics*. With a firm grasp of core concepts of few basic and commonly used tools, one can jump start into the wide field of biology using computation.

We highlight the areas in which computation has forayed in the field of biological sciences. Although every sub-field of biological sciences has been touched by computation, we take a bird's eye view of the areas of interest for the present book.

INFORMATION AGE IN BIOLOGY

Biologists have been collecting, organizing, and interpreting a lot of experimental information for the past few centuries. The field of taxonomy in botany and zoology, is one good example of how the information revolution started in biology. It is very difficult, even for some gifted with very good memory, to memorize the complete taxonomy of a given species of plant or animal.

At the end of the 1800s, *vitalism*—the idea that life had features that were distinct from the rest of the physical world—was still popular. It was the parallel progress in genetics and biochemistry that helped bring *vitalism* to a close. It took until the 1940s for the two fields to overlap, as researchers started to study the genetics of biochemical pathways and the biochemistry of heritable material. This is when the central role of DNA became clear. And that's what inspired Watson and Crick (plus a number of others) to think that understanding the structure of DNA would be critical.

The double helix moved biology into the information age, as Lander put it, since it answered some key questions: how can you store information, faithfully replicate it, but still allow room for the new variants that drive evolution? The identification of the genetic code was the next (and obvious) step in understanding how information was stored in a string of nucleotides. The problem, Lander said, was that you couldn't actually read any of this—at least not as it exists in a cell.

From a biochemical perspective, all DNA was more or less the same, so you couldn't deal with that information by the standard methodology of the time. It took the development of molecular biology and recombinant DNA to start dealing with the actual information content.

"How do you ask how to read DNA?" Lander asked.

"You ask the master—the cell. The cell is in the business of reading the information in DNA."

Molecular biology and biotechnology developed around the purification and use of the proteins used by the cells themselves to manipulate DNA.

The next hurdle, according to Lander, was that we had no idea how to apply the principles of molecular biology to something as big as the human genome in a systematic way. That problem first started to crack with the development of RFLP mapping, when human genes were first associated with a nearby DNA feature that could only be revealed using the techniques of molecular biology. In principle, these worked. But Lander said without the genome sequence, each research group had to essentially start from scratch. For an in-depth description of RFLP mapping, reader is referred to a book hosted at NCBI named **Modern Genetic Analysis**.

Conceptually, he said the key step was the development of a hierarchical map. Lay out genetic markers on a map, identify the DNA associated with those markers, and then dig down into the actual DNA sequences. The first human genetic map appeared in 1987. That set the stage for the genome sequencing to kick off in earnest in the 1990s. The final draft was announced in 2003, on the 50th anniversary of the Watson and Crick paper. Simplified representation is the primary step in comprehending the hidden complexities in genetic code and functioning of proteins. Exploiting the tremendous power of computation, demands that the these building blocks be represented as a single alphabet (Figure 1). This has both merits and demerits. The single letter representation takes away all the feature (chemical, physical and biological), but is the preferred way of representing sequences. Hence, we will be exploring ways of sequence manipulation in this book.

We will explore the concept of information encoding in our DNA in the next section.

DEOXYRIBOSE NUCLEIC ACID (DNA)

DNA stands for ***deoxyribonucleic acid***. Like a protein, a molecule of DNA is a chain of simpler molecules. Actually it is a *double* chain, but let us first understand the structure of one simple chain, called **strand.** It has a backbone consisting of repetitions of the same basic unit. This unit is formed by a sugar molecule called 2'-deoxyribose attached to a phosphate residue. The sugar molecule contains five carbon atoms, and they are labelled 1' through 5' (Figure 2). The bond that creates the backbone is between the 3' carbon of one unit, the phosphate residue, and the 5' carbon of the next unit. For this reason, DNA molecules also have an **orientation**. This by convention starts at 5' end and finishes at 3' end.

Attached to each 1' carbon in the backbone are other molecules called **bases**. There are four kinds of bases: adenine (A), guanine (G), cytosine (C), and thymine (T). Bases A and G belong to a group of substances called *purines,* whereas C and T belong to the *pyrimidines.* When we see the basic unit of a DNA molecule as consisting of the sugar, the phosphate, and its base, we call that a **nucleotide.** Thus, although bases and nucleotides are not the same thing, we can speak of a DNA molecule having 200 bases or 200 nucleotides. A DNA molecule having a few nucleotides is referred to as an *oligonucleotide.* DNA molecules in nature are very long, much longer than proteins. In a human cell, DNA molecules have hundreds of millions of nucleotides.

DNA is double stranded. The two strands are held together by **complementary base pairing**. This was the original contribution of Watson and Crick. In this scheme, A always pairs with T and C with G. This is also known as *Watson-Crick base pairing.*

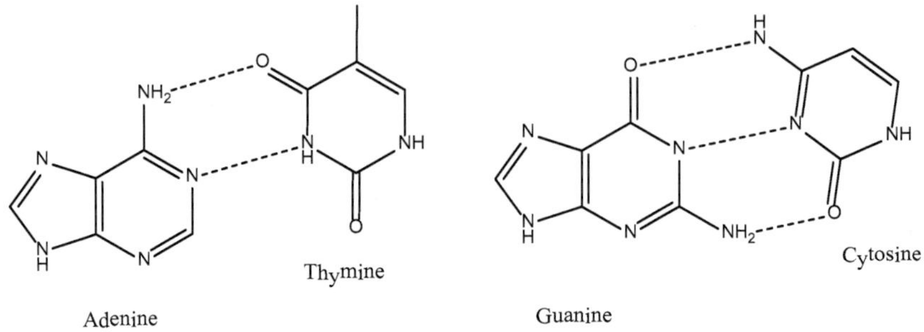

Figure 1.1: Watson-Crick Base pairing.

This base pairing leads to the *anti-parallel strands* making DNA combining to form a **double helix** with bases on the inside of helix and sugar-phosphate forming the backbone of the helix.

Now let us see how information is encoded in DNA.

The importance of DNA molecules is that the information necessary to build each protein or RNA found in an organism is encoded in DNA molecules. For this reason, DNA is sometimes referred to as "the blueprint of life." Each cell of an organism has a few very long DNA molecules. Each such molecule is called a **chromosome**.

- The first important thing to know about this DNA is that certain contiguous stretches along it encode information for building proteins, but others do not.
- The second important thing is that to each different kind of protein in an organism there usually corresponds one and only one contiguous stretch along the DNA. This stretch is known as a **gene.**

Proteins are essentially gene products. A protein, as we have seen, is a chain of amino acids. Therefore, to "specify" a protein all you have to do is to specify each amino acid it contains. And that is precisely what the DNA in a gene does, using triplets of nucleotides to specify each amino acid. Each nucleotide triplet is called a **codon.** The table that gives the correspondence between each possible triplet and each amino acid is the so-called **genetic code**. See the diagram below for understanding which triplets code for which one of the naturally occurring amino acids. The *start codon* AUG codes for Met, and the *stop codons* are UAA, UAG and UGA.

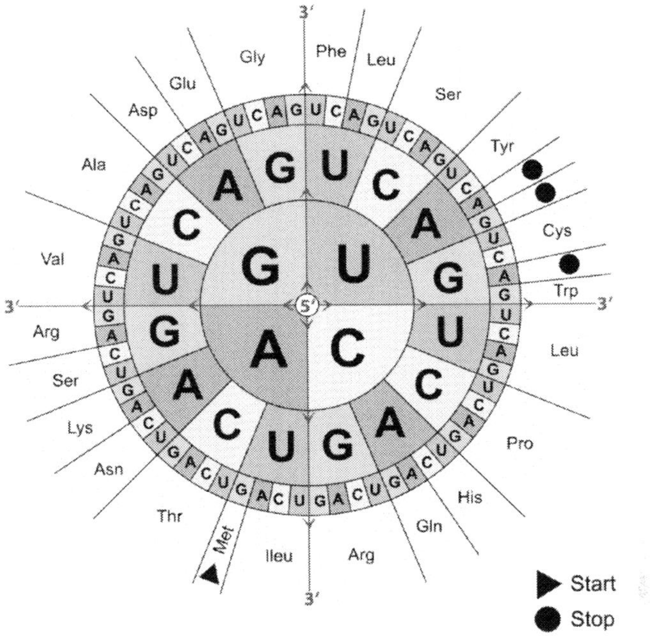

Figure 1.2: Standard translation code.

In this book we will not be focussing on the basics of biology, for which the reader is referred to standard textbooks in the area. Rather, we fill shift our focus on understanding the concepts behind some of the most commonly used methods in biological computation to get meaningful insights into biology.

HUMAN GENOME PROJECT (HGP)

Any book on computational biology is incomplete without due reverence to the multinational efforts to decipher the genetic code of *Homo sapiens*. Draft genome was published by HGP in 2001 and the complete genome in 2003. The main findings of the above studies were:

- Approximately 22000 protein-coding genes in humans.
- Human genome has substantial segmental duplications (repeated sections of DNA).

The genome paved the way for intensive studies on many aspects of human biology. Ranging from understanding gene/protein function to disease implications to evolution, almost every field in biology came under the grip of human genome. The project also inspired genomic works in other areas as well, like plants and agriculture.

MOLECULAR MODELLING

At almost the same time as the publishing of human genome, there was a remarkable rise in the interest in performing molecular modelling and simulation. Scientists have agreed upon one thing, and it is that biological systems are highly *complex*. A *deterministic* approach to understanding biology is that - life is a product of 'upward' causation from genes to cells, organs, systems and whole organism. Then, the question arises where one focus for modelling should? In other words, what should be the starting point. The answer to the question lies in your understanding of the problem in the light of resources accessible to you and the level of expertise in the subject area. Computationally all these can be addressed, at least qualitatively.

It is worthwhile to keep in mind the hunch of Denis Noble -

"*My hunch is that twenty-first century will witness the consummation of the marriage of mathematics and medicine. Computational biology will then simply have been a stepping stone to an even greater transformation of biological science.*"

Molecular modelling essentially focusses on the computational modelling of biomolecules at an atomic level. The field came into existence, once crystal structures of proteins become available. Although modelling of small molecules was in existence, but the methods needed to model bigger molecules had to wait till the structures of the biomolecules like proteins, DNA and RNA became available.

In retrospect, the very first molecular model was the structure of DNA which has become an iconic symbol over a period of six decades. Watson and Crick were able to propose a molecular model of DNA based on the experimental information available in the form of x-ray diffraction patterns.

TOOLS IN INFORMATION AGE

With the data being generated at a rapid pace daily in biology, demands tools which can be used to process the data and comprehend the information. Computers and high-level computer languages are the preferred tools for addressing the problem of information processing in biology and its subfields. We will be directing our focus on understanding the two important languages, **Python** and **R**, which have found wide usage in present day computational biology. There are many other languages like Java, Perl etc., which are also frequently employed to address biological questions computationally.

GROWING DATA

Bioinformaticians are concerned with deriving biological understanding from large amounts of data with specialized skills and tools. Early in biology's history, the datasets were small and manageable. Most biologists could analyze their own data after taking a statistics course, using Microsoft Excel on a personal desktop computer. However, this is all rapidly changing. Large sequencing datasets are widespread, and will only become more common in the future. Analyzing this data takes different tools, new skills, and many computers with large amounts of memory, processing power, and disk space.

In a relatively short period of time, sequencing costs dropped drastically, allowing researchers to utilize sequencing data to help answer important biological questions. Early sequencing was low-throughput and costly. Whole genome sequencing efforts were expensive (the human genome cost around $2.7 billion) and only possible through large collaborative efforts. Since the release of the human genome, sequencing costs have decreased exponentially until about 2008. With the introduction of next-generation sequencing technologies, the cost of sequencing a megabase of DNA dropped even more rapidly. At this crucial point, a technology that was only affordable to large collaborative sequencing efforts (or individual researchers with very deep pockets) became affordable to researchers across all of biology.

The growth of biological databases, on the other hand, is equally impressive. To make matters even more complicated, new tools for analyzing biological data are continually being created, and their underlying algorithms are advancing.

Unfortunately, this abundance and rapid development of bioinformatics tools has serious downsides. Often, bioinformatics tools are not adequately benchmarked, or if they are, they are only benchmarked in one organism. This makes it difficult for new biologists to find and choose the best tool to analyze their data. To make matters more difficult, some bioinformatics programs are not actively developed so that they lose relevance or carry bugs that could negatively affect results. All of this makes choosing an appropriate bioinformatics program in your own research difficult. More importantly, its imperative to critically assess the output of bioinformatics programs run on your own data.

DATA SKILLS

With the nature of biological data changing so rapidly, how are you supposed to learn bioinformatics? With all of the tools out there and more continually being created, how is a biologist supposed to know whether a program will work appropriately on her organisms data?

The solution is to approach bioinformatics as a bioinformatician does: try stuff, and assess the results. In this way, bioinformatics is just about having the skills to experiment with data using a computer and understanding your results.

The limiting factor for most biologists is having the data skills to freely experiment and work with large data on a computer. The goal of this book is to teach you the bioinformatics data skills necessary to allow you to experiment with data on a computer as easily as you would run experiments in the lab.

Unfortunately, many of the biologist's common computational tools can't scale to the size and complexity of modern biological data. Complex data formats, interfacing numerous programs, and assessing software and data make large bioinformatics datasets difficult to work with. Learning core bioinformatics data skills will give you the foundation to learn, apply, and assess any bioinformatics program or analysis method. In 10 years, bioinformaticians may only be using a few of the bioinformatics software programs around today. But we most certainly will be using data skills and experimentation to assess data and methods of the future.

One needs to become comfortable with the command line tools and then gain some expertise in using the packages developed by other experts in the field initially. With improved efficiency and good coding skills one can also develop and publish code, answering a specific biological question.

YOUR COMPUTER

Humans doing rote activities tend to make many mistakes. One of the easiest ways to make your work more robust is to have your computer do as much of this rote work as possible. This approach of automating tasks is more robust because it decreases the odds you'll make a trivial mistake such as accidentally omitting a file or naming output incorrectly.

For example, running a program on 20 different files by individually typing out (or copy and pasting) each command is fragile—the odds of making a careless mistake increase with each file you process. In bioinformatics work, it's good to develop the habit of letting your computer do this sort of repetitive work for you. Instead of pasting the same command 20 times and just changing the input and output files, write a script that does this for you. Not only is this easier and less likely to lead to mistakes, but it also increases reproducibility, as your script serves as a reference of what you did to each of those files. Leveraging the benefits of automating tasks requires a bit of thought in organizing up your projects, data, and code. Simple habits like naming data files in a consistent way that a computer (and not just humans) can understand can greatly facilitate automating tasks and make work much easier.

DON'T REINVENT THE WHEEL - USE EXISTING LIBRARIES

Existing open source libraries have two advantages over libraries you write yourself: a longer history and a wider audience. Both of these advantages translate to fewer bugs. Bugs in software are similar to the proverbial problem of finding a needle in a haystack. If you write your own software library (where a few bugs are bound to be lurking), you're one person looking for a few needles. In contrast, open source software libraries in essence have had many more individuals looking for a much longer time for those needles. Consequently, bugs are more likely to be found, reported, and fixed in these open source libraries than your own home-brewed versions.

An example: Protein structure prediction

The genetic code of an organism is encoded in its DNA. DNA has coding regions which result in the production of the *workhorses* of a cell - *proteins*. Both DNA and proteins are linear polymers of fundamental building blocks. DNA is composed of four **bases** *adenosine* (A) - *thymine* (T) - *cytosine* (C) - *guanosine* (G). Proteins are made of 20 **amino acid residues**. The chemical and molecular structures of the building blocks of DNA and proteins are shown in *Figure 1.1*. Both the DNA and proteins are linear chains of these fundamental building blocks. The linear chains are represented, in shorthand, as one-letter words in one dimension. These 1-dimensional sequences have a definite 3-dimensional (3D) structure. In a protein, amino acids are connected with each other through peptide bonds. Hence, proteins are also known as ***polypeptide chains***.

The "holy grail" of in the field of protein structure is *ab initio* protein structure prediction. Since the demonstration by *Anfinsen* that unfolded proteins can refold to their native three-dimensional conformation, or *tertiary structure*, strictly on the basis of their amino acid sequence, or ***primary structure***, protein chemists have sought computational means of predicting tertiary structure from primary sequence. Despite recent encouraging progress, however, a comprehensive solution to this problem remains elusive. Without delving into the voluminous lore surrounding this field, it is perhaps instructive nonetheless to offer a superficial characterization of the nature of the challenge, which is essentially two-fold.

Despite Anfinsen's result suggesting that the primary sequence plus thermodynamic principles should suffice to completely account for the folded state, the relevant aspects of those principles and the best way to apply them is still not certain. Moreover, it is also the case that there are exceptions to the thesis itself, due to eventualities such as covalent modification, interactions with lipid membranes, the necessity of cofactors, and the transient involvement of additional proteins to assist in folding. This illustrates the first of several principles confounding solutions to many of the grand challenges in computational biology: that *there are no rules without exception in biology* says David Searls.

- The search space of the problem is so daunting, because of the vast range of possible conformations of even relatively short polypeptides, that protein chemists have even been led to wonder how the molecule itself can possibly explore the space adequately on its progress to the folded state. Whatever solution the efficient analog computer of the protein uses to prune this search space, the heuristic is not yet evident in computational approaches to the problem.

One obvious trick is to first solve the problem in some small, tractable subregion, and then to find a way to compose such intermediate results. In this regard, it is discouraging to find that, in the context of entire proteins, identical short stretches of amino acid sequence can exist in completely different conformations. This illustrates a second confounding principle: *biological phenomena invariably have nonlocal components.*

A recurring problem in sequence analysis in general is that, while strictly local effects may dominate any phenomenon (e.g. a consensus binding site), action at a distance always tends to throw a wrench into the works. In the case of protein folding, the reason is immediately evident, since folding tends to bring distant parts of the primary structure into apposition, and it is easy to imagine such interactions "overruling" any local tendency.

However, at times nature can also be benevolent to analysis, and it lends a helping hand in this instance by establishing some highly regular local themes that recur often enough to constitute useful structural clichés. These are the *α-helices* and *β-sheets* that comprise ***secondary structure,*** and the bulk of amino acids in proteins can be identified as participating in one or the other of these characteristic arrangements.

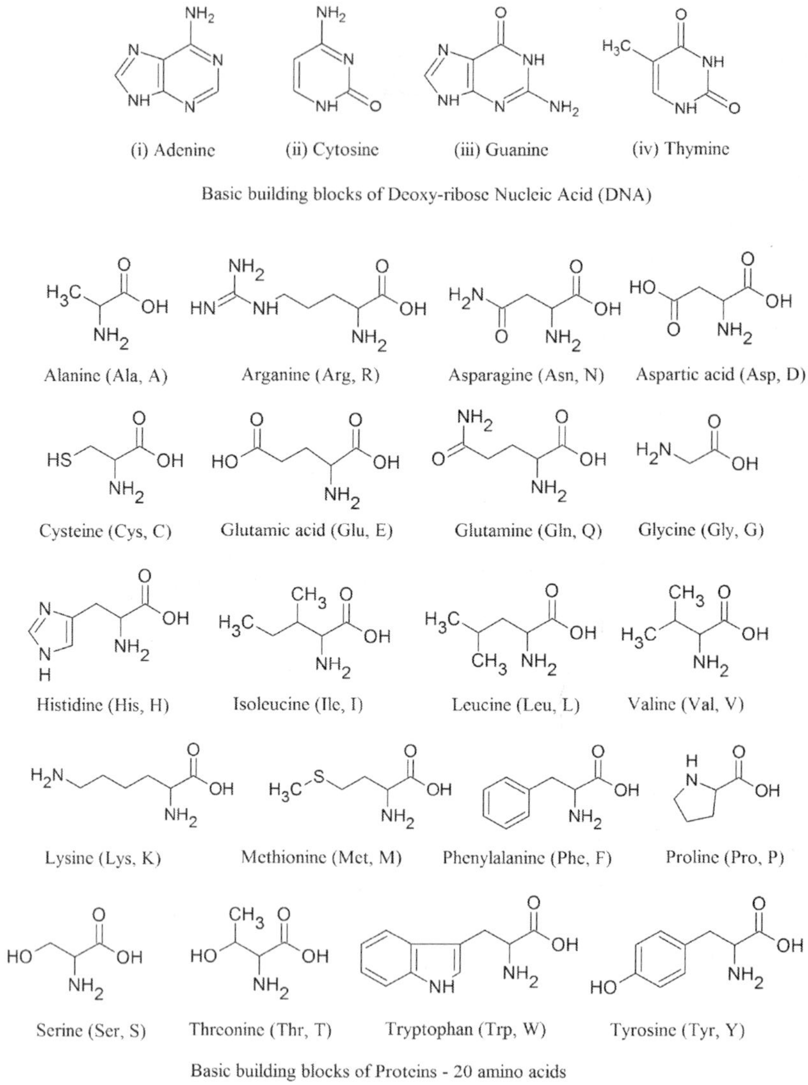

Figure 1.3: Basic chemical structures of bases in DNA and amino acid residues in proteins.

There is a third important class called *β-turns*, which are essentially involved in the reversals of the secondary structures. This results in the characteristic ***three-dimensional structure*** of the protein.

The problem of predicting secondary structure features, which occur in relatively short runs of amino acids, is a minor variant of the protein structure prediction problem, which has been addressed only with decidedly asymptotic success. That is, the earliest empirical systems addressing this problem gave accuracies of 56-60%, which proved not to be terribly useful in practice. Since that time, by a combination of more sophisticated techniques and greater availability of homologous sequences, there has been a significant improvement, yet there is a sense that there is some sort of barrier at around 75% accuracy. In all likelihood this is another manifestation of the nonlocality principle.

Still other simplified variants of the protein folding problem are yielding useful technologies, for example by reversing the problem in cases where the structure of a related protein is known. Techniques of homology modelling and fold recognition ("threading") represent very active areas of research, and in fact have all been part of the **CASP** (**C**ritical **A**ssessment of techniques for protein **S**tructure **P**rediction) contests that have been conducted for several years now to promote objective measurement and progress in the field. That such an institution has been established as a focus for work in this area is itself indicative of the "grand challenge" nature of the overall endeavor.

Before we close this chapter, we will try to summarize the approach which a biological data scientist should adopt when using computational tools. Someone has to make the decisions about what data to collect, and why. That person needs to be formulating questions and hypotheses and making a plan for how the problem will be attacked. And that someone is the data scientist or our beloved data science team.

Let's revise or at least add an overlay to make clear that the data scientist needs to be involved in this process throughout, meaning they are involved in the actual coding as well as in the higher-level process, as shown below.

With this short introduction, we have highlighted the important developments in the field of computational biology. This should inspire you to explore options in the area of biology on computational fronts.

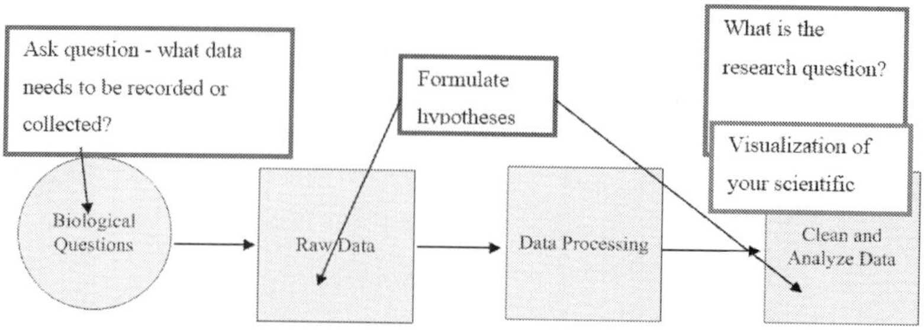

Figure 1.4: Role of a biological data scientist in simplified terms.

THE ROLE OF COMPUTING IN SCIENCE

Science has traditionally been divided into **experimental** and **theoretical** disciplines, but during the last several decades computing has emerged as a very important player in scientific investigations. Scientific computing is closely knit with theory, but it also has many themes in common with experimental work. In most fields of science, computational work is an important complement to both experiments and theory, and nowadays a vast majority of both experimental and theoretical papers involve some numerical calculations, simulations or computer modeling. During the last few decades, since the advent of genomics, computational biology has developed into a field of independent investigation. We now turn our focus on an often-neglected issue in computational studies.

In computational sciences, there are not yet any well-established guidelines for how source code and generated data should be handled. For example, it is relatively rare that source code used in simulations for published papers are provided to readers, in contrast to the open nature of experimental and theoretical work. And it is not uncommon that source code for simulation software is withheld and considered a competitive advantage. However, this issue has recently started to attract increasing attention, and a number of editorials in high-profile journals have called for increased *openness* in computational sciences. Some prestigious journals, including *Science*, have even started to demand authors to provide the source code for simulation software used in publications to readers upon request. Discussions are also ongoing on how to facilitate distribution of scientific software, for example as supplementary materials to scientific papers. This has become norm in computational biology.

REQUIREMENTS ON SCIENTIFIC COMPUTING

Replication and **reproducibility** are two of the cornerstones in the scientific method. With respect to numerical work, complying with these concepts have the following practical implications:

- *Replication*: An author of a scientific paper that involves numerical calculations should be able to rerun the simulations and replicate the results upon request. Other scientist should also be able to perform the same calculations and obtain the same results, given the information about the methods used in a publication.
- *Reproducibility*: The results obtained from numerical simulations should be reproducible with an independent implementation of the method, or using a different method altogether.

To achieve these goals, we need to:

- Keep and take note of *exactly* which source code and version that was used to produce data and figures in published papers.
- Record information of which version of external software that was used. Keep access to the environment that was used.
- Make sure that old codes and notes are backed up and kept for future reference.
- Be ready to give additional information about the methods used, and perhaps also the simulation codes, to an interested reader who requests it (even years after the paper was published!).
- Ideally codes should be published online, to make it easier for other scientists interested in the codes to access it.

TOOLS FOR MANAGING SOURCE CODE

Ensuring replicability and reproducibility of scientific simulations is a *complicated problem*, but there are good tools to help with this:

- Revision Control System (RCS) software.
  - Good choices include:
    - **git** - *http://git-scm.com*
    - **mercurial** - *http://mercurial.selenic.com*. Also known as **hg**.
    - **subversion** - *http://subversion.apache.org*. Also known as **svn**.
- Online repositories for source code. Available as both private and public repositories.
  - Some good alternatives are
    - Github - *http://www.github.com*
    - Bitbucket - *http://www.bitbucket.com*
    - Privately hosted repositories on the university's or department's servers.

Repositories are also excellent for version controlling manuscripts, figures, thesis files, data files, lab logs, etc. Basically, for any digital content that must be preserved and is frequently updated. Again, both public and private repositories are readily available. They are also excellent collaboration tools!

SUMMARY

We have given a panoramic view of computational biology. This should introduce a beginner to the areas of investigation, which are state-of-the-art.

REFERENCES

1. Roger D. Peng. "Reproducible Research in Computational Science", *Science* **334**, 1226 (**2011**).

2. A. Morin et al., Shining Light into Black Boxes, *Science* **336**, 159-160 (**2012**).
3. D.C. Ince, The case for open computer programs, Nature **482**, 485 (**2012**).
4. https://arstechnica.com/science/2012/12/eric-lander-describes-how-biology-entered-the-information-age/
5. http://biobook.kuensting.org/bb/genetics/dna.htm
6. https://www.ncbi.nlm.nih.gov/books/NBK21320/
7. Venter, JC; et al. The sequence of the human genome. *Science*. **291**: 1304–1351 (**2001**).
8. Venter D. A Part of the Human Genome Sequence. *Science*. **299**: 1183–4 (**2003**).
9. Noble D. The Rise of Computational Biology. *Nat. Rev. Mol. Cell. Biol.* **3**: 460-463 (**2002**).

2

PRE-REQUISITES

We introduce some good computational *laboratory practices* in this chapter. We expect that the reader is working on a **Linux** system. Every bioinformatics project begins with an empty project directory, so it's fitting that this book begin with a chapter on **project organization**. In this chapter, we'll look at some of the best practices in organizing your bioinformatics project directories. Without much ado let us get up and start working on the basics of creating an organized directory structure. We have been inspired a great deal from the excellent work presented in the book with the title:

- Bioinformatics Data Skills - Reproducible and Robust Research with Open Source Tools

It is recommended that the interested reader explore the topics covered in the above book. It should help you think and organize your research projects, with a great emphasis on *reproducibility*.

PROJECT DIRECTORIES

Creating a well-organized directory structure is the foundation of a *reproducible* bioinformatics project. The actual process is quite simple:

laying out a project only entails creating a few directories with *mkdir* and empty README files. But this simple initial planning pays off in the long term. For large projects, researchers could spend years working in this directory structure.

All files and directories used in your project should live in a single project directory with a clear name. During the course of a project, you'll have amassed data files, notes, scripts, and so on - if these were scattered all over your hard drive (or worse, across many computers' hard drives), it would be a nightmare to keep track of everything. Even worse, such a disordered project would later make your research nearly impossible to reproduce and comprehend in an extreme scenario. Keeping all of your files in a *single directory* will greatly simplify things for you and your collaborators, and facilitate reproducibility.

First, create a directory (let us call it the *primary* directory) with a meaningful name for the project at hand. Let's say, for example, *mcb* for the present illustration. A nick name for project is not advised. You could also create giving a name like *'methods in computational biology'*. But, the empty spaces in the names could pose a problem. Hence, as a workaround try to add *underscores* (_).

A good and recognizable name for the project directory and subdirectories is a must. For it accords readability and easy comprehension of the data inside the directory and research project. Issue the follow commands to accomplish this objective.

```
> mkdir mcb
> cd mkdir
> mkdir data
> cd data
> mkdir seqs scripts analysis
```

you have made three separate directories under *data* folder for holding sequences, scripts and analyses

Bioinformaticians, software engineers, and system administrators spend *a lot* of time typing in a terminal. It's no surprise these individuals collect tricks to make this process as efficient as possible. As you spend more time in the shell, you'll find investing a little time in learning these tricks can save you a lot of time down the road.

One useful trick is *shell expansion*. Shell expansion is when your shell (e.g., Bash, which is likely the shell you're using) expands text for you so you don't have to type it out. If you've ever typed

> cd ~

to go to your home directory, you've used shell expansion - it's your shell that expands the *tilde* character (~) to the full path to your home directory.

One another scenario is when you type a few characters of say a command or a file name at the terminal and the shells completes it for you. This is usually accomplished by pressing *tab* button after typing a few starting characters.

Bioinformatics projects involve many *subprojects* and *subanalyses*. For example, the quality of raw experimental data should be assessed and poor quality regions removed before running it through bioinformatics tools like aligners or assemblers, among many research objectives.

Let us consider bash before we move into more involved topics. Actually, it would appear as a refresher for those who are already into using computers and have worked on Linux or Unix operating systems.

BASH SHELL

The original Bourne shell distributed with V7 Unix in 1979 became the standard shell for writing shell scripts. The Bourne shell is still found in

/bin/sh

on many commercial Unix systems. It has not changed that much since its initial release, although it has seen modest enhancements over the years.

The Bash shell provides the following features:

- Input/output redirection
- Wildcard characters for filename abbreviation
- Shell variables and options for customizing the environment
- A built-in command set for writing shell programs
- Shell functions, for modularizing tasks within a shell program

- Job control
- Command-line editing (using the command syntax of either vi or Emacs)
- Access to previous commands (command history)
- Integer arithmetic
- Arrays and arithmetic expressions
- Command-name abbreviation (aliasing)
- Upwards compliance with POSIX
- Internationalization facilities
- An arithmetic for loop

The command interpreter for the Bash shell (*bash*) can be invoked as follows:

bash [options] [arguments]

Almost everybody is familiar with common Unix commands like

- ls - used for listing files and directories
- cat - used for reading file contents on screen
- more - does the same as *cat*, but expects a key press to continue to new screen/page
- *vi* and *vim* are text editors which come with default Unix installation.

This book assumes you're familiar with basic topics such as

- what a *terminal* is,
- what the *shell* is,
- the Unix filesystem hierarchy,
- moving about directories,
- file permissions,
- executing commands, and
- Working with a text editor.

Even if you are not, this chapter should serve as a refresher. In this chapter, our focus is on learning how to use core Unix tools to manipulate and explore plain-text data formats. Many formats in bioinformatics are simple tabular *plain-text files* delimited by a character. The most common tabular plain-text file format used in bioinformatics is *tab-*

delimited. This is not an accident: most Unix tools such as **cut** and **awk** treat *tabs* as delimiters by default. Bioinformatics evolved to favor *tab*-delimited formats because of the convenience of working with these files using Unix tools. *Tab*-delimited file formats are also simple to parse with scripting languages like **Python** and **Perl**, and easy to load into **R**. Tabular plain-text data formats are used extensively in computing. The basic format is simple:

- each row (also known as a *record*) is kept on its own line, and
- each column (also known as a *field*) is separated by some delimiter.

There are three flavors you will encounter: *tab*-delimited, *comma*-separated, and variable *space*-delimited.

Of these three formats, tab-delimited is the most commonly used in bioinformatics. File formats such as BED, GTF/GFF, SAM, tabular BLAST output, and VCF are all examples of *tab*-delimited files. Columns of a tab-delimited file are separated by a single tab character (which has the escape code \t). A common convention (but not a standard) is to include metadata on the first few lines of a tab-delimited file. These metadata lines begin with # to differentiate them from the tabular data records. Because tab-delimited files use a tab to delimit columns, tabs in data are not allowed.

Comma-separated values (CSV) is another common format. CSV is similar to tab-delimited, except the delimiter is a *comma* character. While not a common occurrence in bioinformatics, it is possible that the data stored in CSV format contain commas (which would interfere with the ability to parse it). Some variants just don't allow this, while others use quotes around entries that could contain commas. Unfortunately, there's no standard CSV format that defines how to handle this and many other issues with CSV.

Lastly, there are *space*-delimited formats. A few stubborn bioinformatics programs use a variable number of spaces to separate columns. In general, tab-delimited formats and CSV are better choices than space-delimited formats because it's quite common to encounter data containing spaces.

Despite the simplicity of tabular data formats, there's one major common headache: how lines are separated. Linux and OS X use a single linefeed character (with the escape code \n) to separate lines, while Windows uses a DOS-style line separator of a carriage return and a linefeed character (\r\n).

INSPECTING DATA

With raw data in hand, any curious person would like to take a peek into it. Some advanced readers would want to start analyzing the data. All this boils down to using a few simple tools in Unix for getting started data exploration at command line.

Let us look at some useful terminal tools which come in handy for a quick inspection of your data (usually text and delimited data).

Head And Tail

Many files we encounter in bioinformatics are much too long to inspect with *cat* - running *cat* on a file a million lines long would quickly fill your shell with text scrolling far too fast to make sense of. A better option is to take a look at the top of a file with *head*.

The *head* command reads the first few lines of any text given to it as an input and writes them to *standard output* (which, by default, is the display screen).

head's basic syntax is:

head [options] [file(s)]

The square brackets indicate that the enclosed items are optional. By default, *head* returns the first ten lines of each file name that is provided to it. For example, the following will display the first ten lines of the file named *aardvark* (some file with random text in it, used for illustration) in the current directory (i.e., the directory in which the user is currently working):

$ head aardvark

If more than one input file is provided, head will return the first ten lines from each file, precede each set of lines by the name of the file and separate each set of lines by one vertical space. The following is an example of using head with two input files:

$ head aardvark armadillo

If it is desired to obtain some number of lines other than the default ten, the *-n* option can be used followed by an integer indicating the number of lines desired. For example, the above example could be modified to display the first 15 lines from each file:

$ head -n15 aardvark armadillo

-n is a very tolerant option. For example, it is not necessary for the integer to directly follow it without a space in between. Thus, the following command would produce the same result:

$ head -n 15 aardvark armadillo

In fact, the letter *n* does not even need to be used at all. Just the hyphen and the integer (with no intervening space) are sufficient to tell head how many lines to return. Thus, the following would produce the same result as the above commands:

$ head -15 aardvark armadillo

head can also return any desired number of bytes (i.e., a sequence of eight bits and usually long enough to represent a single character) from the start of each file rather than a desired number of lines. This is accomplished using the *-c* option followed by the number of bytes desired. For example, the following would display the first five bytes of each of the two files provided:

$ head -c 5 aardvark anteater

When head counts by bytes, it also includes the *newline character*, which is a *non-printing* (i.e., invisible) character that is designated by a backslash and the letter *n* (i.e., *n*). Thus, for example, if there are three new, blank lines at the start of a file, they will be counted as three characters, along with the *printing characters* (i.e., characters that are visible on the monitor screen or on paper).

The number of bytes or lines can be followed by a multiplier suffix. That is, adding the letter *b* directly after the number of bytes multiplies it by 512, *k* multiplies it by 1024 and *m* multiplies it by 1048576. Thus, the following command would display the first five kilobytes of the file aardvark:

$ head -c5k aardvark

The -c option is less tolerant than the -n option. That is, there is no default number of bytes, and thus some integer must be supplied. Also, the letter *c* cannot be omitted as can the letter *n*, because in such case head would interpret the hyphen and integer combination as the -n option. Thus, for example, the following would produce an error message something like *head: aardvark: invalid number of bytes*:

$ head -c aardvark

If head is used without any options or *arguments* (i.e., file names), it will await input from the keyboard and will successively repeat (i.e., each line will appear twice) on the monitor screen each of the first ten lines typed on the keyboard. If it were desired to repeat some number of lines other than the default ten, then the -n option would be used followed by the integer representing that number of lines (although, again, it is not necessary to include the letter *n*), e.g.,

$ head -n3

As is the case with other command line (i.e., all-text mode) programs in Linux and other Unix-like operating systems, the output from head can *redirected* from the display monitor to a file or printer using the *output redirection operator* (which is represented by a rightward-pointing angular bracket). For example, the following would copy the first 12 lines of the file *Yuriko* to the file *December*:

$ head -n 12 Yuriko > December

If the file named December did not yet exist, the redirection operator would create it; if it already existed, the redirection operator would overwrite it. To avoid erasing data on an existing file, the *append operator* (which is represented by two consecutive rightward pointing angle brackets) could be used to add the output from head to the end of a file with that name if it already existed (or otherwise create a new file with that name), i.e.,

$ head -n 12 Yuriko >> December

The output from other commands can be sent via a *pipe* (represented by the vertical bar character) to head to use as its input. For example, the following sends the output from the *ls* command (which by default lists the names of the files and directories in the current directory) to head, which, in turn, displays the first ten lines of the output that it receives from ls:

$ ls | head

This output could easily be redirected, for example to the end of a file named *file1* as follows:

$ ls | head >> file1

It could also be piped to one or more filters for additional processing. For example, the *sort* filter could be used with its *-r* option to sort the output in reverse alphabetic order prior to appending file1:

$ ls | head | sort -r >> file1

The *-q* (i.e., *quiet*) option causes head to not show the file name before each set of lines in its output and to eliminate the vertical space between each set of lines when there are multiple input sources. Its opposite, the *-v* (i.e., *verbose*) option, causes head to provide the file name even if there is just a single input file.

The *tail* command is similar to the head command except that it reads the final lines in files rather than the first lines.

As is the case with other commands on Unix-like operating systems, additional information can be obtained about *head* and *tail* by using the *man* and *info* commands to reference the built-in documentation, for example

$ man head

or

$ info tail

less

less is also a useful program for a inspecting files and the output of pipes. *less* is a *terminal pager*, a program that allows us to view large amounts of text in our terminals. Normally, if we *cat* a long file to screen, the text flashes by in an instant - *less* allows us to view and scroll through long files and standard output a screen at a time. Other applications can call the default terminal pager to handle displaying large amounts of output; this is how git log displays an entire Git repository's commit history. You might run across another common, but older terminal pager called more, but less has more features and is generally preferred (the name of less is a play on "less is more").

less runs more like an application than a command: once we start less, it will stay open until we quit it.

$ less some_file.fastq

This will open up the program less in your terminal and display a FASTQ file full of sequences. First, if you need to quit less, press *q*. At any time, you can bring up a help page of all of less's commands with *h*.

Moving around in less is simple: press space to go down a page, and *b* to go up a page. You can use *j* and *k* to go down and up a line at a time (these are the same keys that the editor Vim uses to move down and up). To go back up to the top of a file, enter *g*; to go to the bottom of a file, press *G*. We'll talk a bit about how this works when less is taking input from another program through a pipe in a bit. Let us see some commands which are frequently used with less in Table below.

Table 2.1: *less* commands

Key board button press	Result
spacebar	Next page
b	Previous page
g	First line
G	Last line
j	Down (a line at a time)
K	Up (a line at a time)
/<pattern>	Search down for string
?<pattern>	Search up for string
n	Repeat last search downward
N	Repeat last search upward

One of the most useful features of less is that it allows you to search text and highlights matches. Visually highlighting matches can be an extremely useful way to find potential problems in data.

Wc And Ls

In addition to peeking at a file with *head*, *tail*, or *less*, we may want other bits of summary information about a plain-text data file like the number of rows or columns. With plain-text data formats like tab-delimited and CSV files, the number of rows is usually the number of lines. We can retrieve this with the program *wc* (for word count). The *wc* (i.e., word count) command by default counts the number of lines, words and characters in text.

wc defines a word as a set of contiguous letters, numbers and/or symbols which are separated from other characters by one or more spaces, tabs and/or newline characters (which are generated when the RETURN key is pressed). When counting the number of characters, all characters are counted, not only letters, numbers and symbols, but also spaces, tabs and newline characters. A line is only counted if it ends with a newline character.

wc's syntax is

wc [options] [file_name(s)]

The items in square brackets are optional. If no file names are provided, wc reads from its standard input, which by default is text entered at the keyboard. This can be seen by typing

$ wc

at the command line (i.e., in the all-text mode), pressing the ENTER key to move to a new line and then typing some text on one or more lines. The command is executed (i.e., run) by pressing the ENTER key again and then pressing the CONTROL and d keys simultaneously. This causes wc to write in a new line (under the lines of text) its count of the numbers of lines, words and characters in the text.

The following command counts the number of lines, words and characters in a file named *file1* that resides in the current directory (i.e., the directory in which the user is currently working) and likewise writes them, followed by the name of the file, to standard output, which is by default the display monitor (this is for illustration):

$ wc file1

wc can provide its output for multiple files by listing the name of each separated by a space. For example,

$ wc file1 file2 file3

The numbers of lines, words and characters for each file along with its name will be displayed on a separate line and in the order that the files are listed as arguments (i.e., input files). In the case of multiple arguments such as this, wc also provides an additional line that shows the total number of lines, words and characters for all the files.

Likewise, *wc* can provide a count for all of the text files within a directory. This is accomplished by using the star wildcard character, which represents everything and is designated by an asterisk (*). For example, the following will display the number of lines, words and characters for each file in the current directory (which is represented by a dot) as well as totals for all files in the directory:

```
$ wc . *
```

wc has only a few options, the most commonly used of which restrict the information it provides. The *-l* option tells wc to count only the number of lines, the *-w* option tells it to count only the number of words, the *-m* option tells it to count only the number of characters and the *-c* option tells wc to count only the number of bytes.

Thus, for example, the following displays just the number of words in a file named file4:

```
$ wc -w file4
```

The following displays the number of characters in the same file:

```
$ wc -m file4
```

As is generally the case with commands in Unix-like operating systems, any combination of options can be used together. For example, the following would count both the numbers of lines and words in a file named file5:

```
$ wc -lw file5
```

Redirection can be used with *wc* to create more complex commands. For example, the output from the above command can be redirected using the standard output redirection operator (which is designated by a rightward pointing angle bracket) from the display screen to a file named file6 with the following:

```
$ wc -lw file5 > file6
```

If file6 already exists, its contents will be overwritten; if it does not exist, it will be created. The contents of file6 can be easily confirmed with a text editor or with a command such as cat, which is commonly used to read text files, i.e.,

```
$ cat file6
```

As another example, *wc* with its *-l* option can be used to count the total number of objects (i.e., files, links and directories) in the current directory. This is accomplished by first employing the *ls* command, which by default lists the contents of the current directory, and then sending its output via a pipe (designated by the vertical bar character) to *wc*, which counts the number of lines in the output from ls, i.e.,

```
$ ls | wc -l
```

wc can count the words, lines and/or characters only in text files. However, when used with its *-c* option, it can count the bytes in any type of file, i.e. a plain text file or a binary file. (A binary file is any file that contains at least some non-text data, such as an image file, a compressed text file or an executable program.) This can be a convenient way of determining the size of a file.

For example, the following command displays the size for each *jpeg* image file in the current directory as well as the total for all of them:

```
$ wc -c *.jpg
```

You can try this command on your FASTA, FASTQ GTF or BED files.

TEXT PROCESSING WITH AWK

We'll introduce the basics of *awk* in this section - enough to get you started with using *awk* in bioinformatics. While *awk* is a full-fledged programming language, it's a lot less expressive, terse and less powerful than **Python**. If you need to implement something complex, it's likely better (and easier) to do so in **Python**. The key to using *awk* effectively is to reserve it for the subset of tasks it's best at:

quick data-processing tasks on tabular data

Learning Awk also prepares us to learn *bioawk*. To learn awk, we'll cover two key parts of the awk language: how awk processes records, and pattern-action pairs. After understanding these two key parts the rest of the language is quite simple.

First, awk processes input data a *record* at a time. Each record is composed of *fields*. The separate chunks that awk automatically separates are known as fields. Because awk was designed to work with tabular data, each *record is a line*, and each *field is a column's entry* for that record. Awk automatically assigns the entire record to the variable $0, and field one's value is assigned to $1, field two's value is assigned to $2, field three's value is assigned to $3, and so forth. Second, we build Awk programs using one or more of the following structures:

pattern {action}

If the pattern's expression evaluates to **true** or the regular expression *matches*, the statements inside *action* are run. In awk these are known as *pattern-action* pairs and we can chain multiple *pattern-action pairs* together (these are usually separated by semicolons).

- If we **omit** the *pattern*, awk will run the action on all records.
- If we **omit** the *action* but specify a pattern, awk will print all records that match the pattern.

This simple structure makes awk an excellent choice for quick text-processing tasks. This is a lot to take in, but these two basic concepts - *records* and *fields*, and *pattern-action pairs* - are the foundation of writing text-processing programs with awk. Let's see some examples. First, we can simply mimic *cat* by omitting a pattern and printing an entire *record* with the variable $0:

awk '{print $0}' some.bed

print prints a string. Optionally, we could omit the $0, because print called without an argument would print the current record. Awk can also mimic **cut**:

awk '{ print $2 "\t" $3 }' example.bed

Here, we're making use of Awk's string **concatenation**. Two strings are concatenated if they are placed next to each other with no argument. So for each record, $2"\t"$3 concatenates the second field, a tab character, and the third field. This is far more typing than cut -f2,3, but demonstrates how we can access a certain column's value for the current record with the numbered variables $1, $2, $3, etc.

Let's now look at how we can incorporate simple pattern matching. Suppose we wanted to write a filter that only output lines where the length of the feature ® (end position - start position) > 9.

Awk supports arithmetic with the standard operators +, -, *, /, % (remainder), and ^ (exponentiation). We can subtract within a pattern to calculate the length of a feature, and filter on that expression:

awk '$3 - $2 > 9' some.bed

We provide comparison and logical operators below:

Table 2.2: A table showing comparison and logical operators available in *awk*.

Comparison	Description
a == b	a is equal to b
a != b	a is not equal to b
a < b	a is less than b
a > b	a is greater than b
a <= b	a is less than or equal to b
a >= b	a is greater than or equal to b
a ~ b	a matches regular expression pattern b
a !~ b	a does not match regular expression pattern b
a && b	logical and a and b
a || b	logical or a and b
!a	not a (logical negation)

We leave the reader with this short introduction to further explore the *magic* of command line tools in pre-processing biological data. One good starting point is the book by Vince Buffalo.

SUMMARY

In this chapter we have presented some prerequisites for embarking on any computational project. The reader is expected to get comfortable with these and many additional command line tools for a fluent handling of biological text data.

REFERENCES

1. Bioinformatics Data Skills by Vince Buffalo, O'Reilly Media Inc. (2015)

3

SCIENTIFIC COMPUTING IN PYTHON

INTRODUCING PYTHON

Python is a modern, general-purpose, object-oriented, high-level programming language. Many books are available which introduce a reader to Python. We also venture in this chapter to introduce Python from the perspective of scientific computing. Let us first have a look at some of the general characteristics of Python:

- **Clean and simple language:** Easy-to-read and intuitive code, easy-to-learn minimalistic syntax, maintainability scales well with size of projects.
- **Expressive language:** Fewer lines of code, fewer bugs, easier to maintain.

Before we delve into Python, some technical details are provided below:

- **Dynamically typed:** No need to define the *type* of variables, function *arguments* or return types.
- **Automatic memory management:** No need to explicitly *allocate* and *deallocate* memory for variables and data arrays. No memory leak bugs.

- **Interpreted:** No need to *compile* the code. The Python interpreter reads and executes the python code directly.

You may ask what the advantages of coding in Python are. Let us enumerate them below:

- The main advantage is *ease of programming*, minimizing the time required to develop, debug and maintain the code.
- Well-designed language that encourage many *good programming practices*.
- *Modular* and object-*oriented programming*, good system for packaging and re-use of code. This often results in more transparent, maintainable and bug-free code.
- *Documentation* tightly integrated with the code.
- A large *standard library*, and a large collection of add-on packages.

Let us also pay some attention to some of the disadvantages:

- Since Python is an interpreted and dynamically typed programming language, the *execution* of python code can be *slow* compared to compiled statically typed programming languages, such as **C** and **Fortran**.
- Somewhat *decentralized*, with different environment, packages and documentation spread out at different places. Can make it harder to get started.

WHAT MAKES PYTHON SUITABLE FOR SCIENTIFIC COMPUTING?

This is the most obvious question that comes up to a curious reader. Take the information below on word. You will realize the power as you move up the ladder in Python learning!

- **Python** has a strong position in scientific computing:
  - Large community of users, easy to find help and documentation.
- Extensive ecosystem of scientific libraries and environments

  - *NumPy*: http://numpy.scipy.org - Numerical Python
  - *scipy*: http://www.scipy.org - Scientific Python
  - *matplotlib*: http://www.matplotlib.org - graphics library
- Great performance due to close integration with time-tested and highly optimized codes written in **C** and **Fortran**:
  - blas, atlas blas, lapack, arpack, Intel MKL, ...
- Good support for
  - Parallel processing with processes and threads
  - Interprocess communication (MPI)
  - GPU computing (OpenCL and CUDA)
- Readily available and suitable for use on high-performance computing clusters.
- No license costs, no unnecessary use of research budget.

Python Environments

Python is not only a programming language, but often also refers to the standard implementation of the interpreter (technically referred to as CPython) that actually runs the python code on a computer. There are also many different environments through which the python interpreter can be used. Each environment has different advantages and is suitable for different workflows. One strength of python is that it is versatile and can be used in complementary ways, but it can be confusing for beginners so we will start with a brief survey of python environments that are useful for scientific computing.

Python Interpreter

The standard way to use the Python programming language is to use the Python interpreter to run python code. The python interpreter is a program that reads and execute the python code in files passed to it as arguments. At the command prompt, the command python is used to invoke the Python interpreter. For example, to run a file my-program.py that contains python code from the command prompt, use:

$ python my-program.py

We can also start the interpreter by simply typing python at the command line, and interactively type python code into the interpreter. This is often how we want to work when developing scientific applications, or when doing small calculations. But the standard python interpreter is not very convenient for this kind of work, due to a number of limitations.

Ipython

IPython is an interactive shell that addresses the limitation of the standard python interpreter, and it is a work-horse for scientific use of python. It provides an interactive prompt to the python interpreter with a greatly improved user-friendliness.

Some of the many useful features of IPython includes:

- Command history, which can be browsed with the up and down arrows on the keyboard.
- Tab auto-completion.
- In-line editing of code.
- Object introspection, and automatic extract of documentation strings from python objects like classes and functions.
- Good interaction with operating system shell.
- Support for multiple parallel back-end processes, that can run on computing clusters or cloud services like Amazon EC2.

Ipython Notebook

IPython notebook is an HTML-based notebook environment for Python, similar to Mathematica or Maple. It is based on the IPython shell, but provides a cell-based environment with great interactivity, where calculations can be organized and documented in a structured way.

Although using a web browser as graphical interface, IPython notebooks are usually run locally, from the same computer that run the browser. To start a new IPython notebook session, run the following command:

$ ipython notebook # now-a-days it is superseded with jupyter notebook

from a directory where you want the notebooks to be stored. This will open a new browser window (or a new tab in an existing window) with an index page where existing notebooks are shown and from which new notebooks can be created.

Spyder

Spyder is a MATLAB-like IDE for scientific computing with python. It has the many advantages of a traditional IDE environment, for example that everything from code editing, execution and debugging is carried out in a single environment, and work on different calculations can be organized as projects in the IDE environment.

Some advantages of Spyder:

- Powerful code editor, with syntax high-lighting, dynamic code introspection and integration with the python debugger.
- Variable explorer, IPython command prompt.
- Integrated documentation and help.

Versions of Python

There are currently two versions of python: Python 2 and Python 3. Python 3 will eventually supercede Python 2, but it is **not** *backward-compatible* with Python 2. A lot of existing python code and packages has been written for Python 2, and it is still the most wide-spread version.

To see which version of Python you have, run

$ python –version

Python 2.7.3

$ python3.2 –version

Python 3.2.3

Several versions of Python can be installed in parallel, as shown above.

We will not cover installation, as it is covered in many places. But, we recommend workingin Anaconda environment.

PYTHON PROGRAM FILES

- Python code is usually stored in text files with the file ending ".py":

myprogram.py

- Every line in a Python program file is assumed to be a Python statement, or part thereof.
  - The only exception is comment lines, which start with the character # (optionally preceded by an arbitrary number of white-space characters, i.e., tabs or spaces). Comment lines are usually ignored by the Python interpreter.
- To run our Python program from the command line we use:

$ python myprogram.py

- On UNIX systems it is common to define the path to the interpreter on the first line of the program (note that this is a comment line as far as the Python interpreter is concerned):

#!/usr/bin/env python

If we do, and if we additionally set the file script to be executable, we can run the program like this:

$ myprogram.py

Character Encoding

The standard character encoding is ASCII, but we can use any other encoding, for example UTF-8. To specify that UTF-8 is used we include the special line at the top of the file.

-*- coding: UTF-8 -*-

cat scripts/hello-world-in-swedish.py

#!/usr/bin/env python

-*- coding: UTF-8 -*-

print("Hej världen!")

!python scripts/hello-world-in-swedish.py

Hej världen!

Other than these two *optional* lines in the beginning of a Python code file, no additional code is required for initializing a program.

Ipython Notebooks

An IPython notebook is stored as a file in the JSON format. The advantage is that we can mix formatted text, Python code and code output. It requires the IPython notebook server to run it though, and therefore isn't a stand-alone Python program as described above. Other than that, there is no difference between the Python code that goes into a program file or an IPython notebook.

Modules

Most of the functionality in Python is provided by *modules*. The Python Standard Library is a large collection of modules that provides *cross-platform* implementations of common facilities such as access to the operating system, file I/O, string management, network communication, and much more.

To use a module in a Python program it first has to be imported. A module can be imported using the **import** statement. For example, to import the module **math**, which contains many standard mathematical functions, we can do:

```
import math
```

This includes the whole module and makes it available for use later in the program. For example, we can do:

```
import math
x = math.cos(2 * math.pi)
print(x)
1.0
```

Alternatively, we can chose to import all symbols (functions and variables) in a module to the current namespace (so that we don't need to use the prefix "math." every time we use something from the math module:

```
from math import *
x = cos(2 * pi)
print(x)
1.0
```

This pattern can be very convenient, but in large programs that include many modules it is often a good idea to keep the symbols from each module in their own namespaces, by using the import math pattern. This would elminate potentially confusing problems with name space collisions.

As a third alternative, we can chose to import only a few selected symbols from a module by explicitly listing which ones we want to import instead of using the wildcard character *:

```
from math import cos, pi
x = cos(2 * pi)
print(x)
1.0
```

Once a module is imported, we can list the symbols it provides using the dir function:

```
import math
print(dir(math))
['__doc__', '__file__', '__name__', '__package__', 'acos', 'acosh', 'asin', 'asinh', 'atan', 'atan2', 'atanh', 'ceil', 'copysign', 'cos', 'cosh', 'degrees', 'e', 'erf', 'erfc', 'exp', 'expm1', 'fabs', 'factorial', 'floor', 'fmod', 'frexp', 'fsum', 'gamma', 'hypot', 'isinf', 'isnan', 'ldexp', 'lgamma', 'log', 'log10', 'log1p', 'modf', 'pi', 'pow', 'radians', 'sin', 'sinh', 'sqrt', 'tan', 'tanh', 'trunc']
```

And using the function help we can get a description of each function (almost .. not all functions have docstrings, as they are technically called, but the vast majority of functions are documented this way).

```
help(math.log)
```

Help on built-in function log in module math:

```
log(...)
   log(x[, base])
```

Return the logarithm of x to the given base.

If the base not specified, returns the natural logarithm (base e) of x.

log(10)

2.302585092994046

log(10, 2)

3.3219280948873626

We can also use the help function directly on modules: Try

help(math)

Some very useful modules form the Python standard library are os, sys, math, shutil, re, subprocess, multiprocessing, threading.

A complete lists of standard modules for Python 2 and Python 3 are available at:

http://docs.python.org/2/library/ and

http://docs.python.org/3/library/, respectively.

PYTHON BASICS

Variables and Types

Variable names in Python can contain alphanumerical characters a-z, A-Z, 0-9 and some special characters such as _. Normal variable names must start with a letter. By convention, variable names start with a lower-case letter, and Class names start with a capital letter. In addition, there are a number of Python keywords that cannot be used as variable names. These keywords are:

and, as, assert, break, class, continue, def, del, elif, else, except,

exec, finally, for, from, global, if, import, in, is, lambda, not, or,

pass, print, raise, return, try, while, with, yield

Note: Be aware of the keyword lambda, which could easily be a natural variable name in a scientific program. But being a keyword, it cannot be used as a variable name.

Fundamental Types

```
# integers
x = 1
type(x)
int
# float
x = 1.0
type(x)
float
# Boolean
b1 = True
b2 = False
type(b1)
bool
# complex numbers: note the use of `j` to specify the imaginary part
x = 1.0 - 1.0j
type(x)
complex
print(x)
(1-1j)
print(x.real, x.imag)
(1.0, -1.0)
```

Type Utility Functions

The module types contain a number of type name definitions that can be used to test if variables are of certain types:

```
import types
# print all types defined in the `types` module
print(dir(types))
['BooleanType', 'BufferType', 'BuiltinFunctionType', 'BuiltinMethodType', 'ClassType', 'CodeType', 'ComplexType',
```

'DictProxyType', 'DictType', 'DictionaryType', 'EllipsisType', 'FileType', 'FloatType', 'FrameType', 'FunctionType', 'GeneratorType', 'GetSetDescriptorType', 'InstanceType', 'IntType', 'LambdaType', 'ListType', 'LongType', 'MemberDescriptorType', 'MethodType', 'ModuleType', 'NoneType', 'NotImplementedType', 'ObjectType', 'SliceType', 'StringType', 'StringTypes', 'TracebackType', 'TupleType', 'TypeType', 'UnboundMethodType', 'UnicodeType', 'XRangeType', '__all__', '__builtins__', '__doc__', '__file__', '__name__', '__package__']

x = 1.0

check if the variable x is a float

type(x) **is** float

True

check if the variable x is an int

type(x) **is** int

False

We can also use the isinstance method for testing types of variables:

isinstance(x, float)

True

Operators and Comparisons

Most operators and comparisons in Python work as one would expect:

- Arithmetic operators +, -, *, /, // (integer division), '**' power

1 + 2, 1 - 2, 1 * 2, 1 / 2

(3, -1, 2, 0)

1.0 + 2.0, 1.0 - 2.0, 1.0 * 2.0, 1.0 / 2.0

(3.0, -1.0, 2.0, 0.5)

Integer division of float numbers

3.0 // 2.0

1.0

Note! The power operators in python isn't ^, but **

2 ** 2

4

Note: The / operator always performs a floating-point division in Python 3.x. This is not true in Python 2.x, where the result of / is always an integer if the operands are integers. to be more specific, 1/2 = 0.5 (float) in Python 3.x, and 1/2 = 0 (int) in Python 2.x (but 1.0/2 = 0.5 in Python 2.x).

- The boolean operators are spelled out as the words and, not, or.

True **and** False

False

not False

True

True **or** False

True

- Comparison operators >, <, >= (greater or equal), <= (less or equal), == equality, is identical.

2 > 1, 2 < 1

(True, False)

2 > 2, 2 < 2

(False, False)

2 >= 2, 2 <= 2

(True, True)

equality

[1,2] == [1,2]

True

objects identical?

l1 = l2 = [1,2]

l1 **is** l2

True

Compound Types: Strings, List and Dictionaries

Strings are the variable type that is used for storing text messages.

```
s = “Hello world”
type(s)
str
# length of the string: the number of characters
len(s)
11
# replace a substring in a string with something else
s2 = s.replace(“world”, “test”)
print(s2)
Hello test
```

We can index a character in a string using []:

```
s[0]
‘H’
```

We can extract a part of a string using the syntax [start:stop], which extracts characters between index start and stop -1 (the character at index stop is not included):

```
s[0:5]
‘Hello’
s[4:5]
‘o’
```

If we omit either (or both) of start or stop from [start:stop], the default is the beginning and the end of the string, respectively:

```
s[:5]
‘Hello’
s[6:]
‘world’
s[:]
‘Hello world’
```

We can also define the step size using the syntax [start:end:step] (the default value for step is 1, as we saw above):

s[::1]

'Hello world'

s[::2]

'Hlowrd'

This technique is called *slicing*. Python has a very rich set of functions for text processing. See for example for more information.

http://docs.python.org/2/library/string.html

String Formatting Examples

print("str1", "str2", "str3") # The print statement concatenates strings with a space

('str1', 'str2', 'str3')

print("str1", 1.0, False, -1j) # The print statements converts all arguments to strings

('str1', 1.0, False, -1j)

print("str1" + "str2" + "str3") # strings added with + are concatenated without space

str1str2str3

print("value = %f" % 1.0) # we can use C-style string formatting

value = 1.000000

this formatting creates a string

s2 = "value1 = %.2f. value2 = %d" % (3.1415, 1.5)

print(s2)

value1 = 3.14. value2 = 1

alternative, more intuitive way of formatting a string

s3 = 'value1 = {0}, value2 = {1}'.format(3.1415, 1.5)

print(s3)

value1 = 3.1415, value2 = 1.5

List

Lists are very similar to strings, except that each element can be of any type. The syntax for creating lists in Python is [...]:

```
l = [1,2,3,4]
print(type(l))
print(l)
<type 'list'>
[1, 2, 3, 4]
```

We can use the same slicing techniques to manipulate lists as we could use on strings:

```
print(l)
print(l[1:3])
print(l[::2])
[1, 2, 3, 4]
[2, 3]
[1, 3]
```

Elements in a list do not all have to be of the same type:

```
l = [1, 'a', 1.0, 1-1j]
print(l)
[1, 'a', 1.0, (1-1j)]
```

Python lists can be inhomogeneous and arbitrarily nested:

```
nested_list = [1, [2, [3, [4, [5]]]]]
nested_list
[1, [2, [3, [4, [5]]]]]
```

Lists play a very important role in Python. For example, they are used in loops and other flow control structures (discussed below). There are a number of convenient functions for generating lists of various types, for example the range function:

```
start = 10
stop = 30
step = 2
```

```
range(start, stop, step)
[10, 12, 14, 16, 18, 20, 22, 24, 26, 28]
# in python 3 range generates an interator, which can be converted to a list using 'list(...)'.
# It has no effect in python 2
list(range(start, stop, step))
[10, 12, 14, 16, 18, 20, 22, 24, 26, 28]
list(range(-10, 10))
[-10, -9, -8, -7, -6, -5, -4, -3, -2, -1, 0, 1, 2, 3, 4, 5, 6, 7, 8, 9]
s
'Hello world'
# convert a string to a list by type casting:
s2 = list(s)
s2
['H', 'e', 'l', 'l', 'o', ' ', 'w', 'o', 'r', 'l', 'd']
# sorting lists
s2.sort()print(s2)
[' ', 'H', 'd', 'e', 'l', 'l', 'l', 'o', 'o', 'r', 'w']
```

Adding, Inserting, Modifying, and Removing Elements from Lists

```
# create a new empty list
l = []
# add an elements using `append`
l.append("A")
l.append("d")
l.append("d")
print(l)
['A', 'd', 'd']
```

We can modify lists by assigning new values to elements in the list. In technical jargon, lists are *mutable*.

```
l[1] = "p"
l[2] = "p"
print(l)
['A', 'p', 'p']
l[1:3] = ["d", "d"]
print(l)
['A', 'd', 'd']
```

Insert an element at a specific index using insert

```
l.insert(0, "i")
l.insert(1, "n")
l.insert(2, "s")
l.insert(3, "e")
l.insert(4, "r")
l.insert(5, "t")
print(l)
['i', 'n', 's', 'e', 'r', 't', 'A', 'd', 'd']
```

Remove first element with specific value using 'remove'

```
l.remove("A")
print(l)
['i', 'n', 's', 'e', 'r', 't', 'd', 'd']
```

Remove an element at a specific location using del:

```
del l[7]
del l[6]
print(l)
['i', 'n', 's', 'e', 'r', 't']
```

See help(list) for more details, or read the online documentation

Tuples

Tuples are like lists, except that they cannot be modified once created, that is they are *immutable.*

In Python, tuples are created using the syntax (..., ..., ...), or even ..., ...:

```
point = (10, 20)
print(point, type(point))
((10, 20), <type 'tuple'>)
```

```
point = 10, 20
print(point, type(point))
((10, 20), <type 'tuple'>)
```

We can unpack a tuple by assigning it to a comma-separated list of variables:

```
x, y = point
print("x =", x)
print("y =", y)
('x =', 10)
('y =', 20)
```

If we try to assign a new value to an element in a tuple we get an error:

```
point[0] = 20
---------------------------------------------------------------------------
TypeError                         Traceback (most recent call last)
<ipython-input-83-ac1c641a5dca> in <module>()
----> 1 point[0] = 20
TypeError: 'tuple' object does not support item assignment
```

Dictionaries

Dictionaries are also like lists, except that each element is a key-value pair. The syntax for dictionaries is {key1 : value1, ...}:

```
params = {"parameter1" : 1.0,
```

```
        "parameter2" : 2.0,
        "parameter3" : 3.0,}
print(type(params))
print(params)
<type 'dict'>
{'parameter1': 1.0, 'parameter3': 3.0, 'parameter2': 2.0}
print("parameter1 = " + str(params["parameter1"]))
print("parameter2 = " + str(params["parameter2"]))
print("parameter3 = " + str(params["parameter3"]))
parameter1 = 1.0
parameter2 = 2.0
parameter3 = 3.0
params["parameter1"] = "A"
params["parameter2"] = "B"
# add a new entry
params["parameter4"] = "D"
print("parameter1 = " + str(params["parameter1"]))
print("parameter2 = " + str(params["parameter2"]))
print("parameter3 = " + str(params["parameter3"]))
print("parameter4 = " + str(params["parameter4"]))
parameter1 = A
parameter2 = B
parameter3 = 3.0
parameter4 = D
```

Control Flow

The Python syntax for conditional execution of code uses the keywords if, elif (else if), else:

```
statement1 = False
statement2 = False
```

```
if statement1:
    print("statement1 is True")
elif statement2:
    print("statement2 is True")
else:
    print("statement1 and statement2 are False")
statement1 and statement2 are False
```

For the first time, here we encountered a peculiar and unusual aspect of the Python programming language: Program blocks are defined by their indentation level.

Loops

In Python, loops can be programmed in a number of different ways. The most common is the **for** loop, which is used together with iterable objects, such as lists. The basic syntax is:

```
for x in [1,2,3]:
    print(x)
1
2
3
```

The for loop iterates over the elements of the supplied list, and executes the containing block once for each element. Any kind of list can be used in the for loop. For example:

```
for x in range(4): # by default range start at 0
    print(x)
0
1
2
3
```

Note: range(4) does not include 4 !

```
for x in range(-3,3):
```

```
    print(x)
-3
-2
-1
0
1
2
for word in ["scientific", "computing", "with", "python"]:
    print(word)
scientific
computing
with
python
```

To iterate over key-value pairs of a dictionary:

```
for key, value in params.items():
    print(key + " = " + str(value))
parameter4 = D
parameter1 = A
parameter3 = 3.0
parameter2 = B
```

Sometimes it is useful to have access to the indices of the values when iterating over a list. We can use the *enumerate* function for this:

```
for idx, x in enumerate(range(-3,3)):
    print(idx, x)
(0, -3)
(1, -2)
(2, -1)
(3, 0)
(4, 1)
(5, 2)
```

List comprehensions: creating lists using for loops:

A convenient and compact way to initialize lists:

```
l1 = [x**2 for x in range(0,5)]
print(l1)
[0, 1, 4, 9, 16]
```

While loops are also commonly employed in programming. Some examples illustrate the usage:

```
i = 0
while i < 5:
    print(i)
    i = i + 1
print("done")
0
1
2
3
4
done
```

Note that the print("done") statement is not part of the while loop body because of the difference in indentation.

Functions

A function in **Python** is defined using the keyword def, followed by a function name, a signature within parentheses (), and a colon :. The following code, with one additional level of indentation, is the function body.

```
def func0():
    print("test")
func0()
test
```

Functions that returns a value use the return keyword:

```
def square(x):
    """
    Return the square of x.
    """
    return x ** 2
square(4)
16
```

We can return multiple values from a function using tuples (see above):

```
def powers(x):
    """
    Return a few powers of x.
    """
    return x ** 2, x ** 3, x ** 4
powers(3)
(9, 27, 81)
x2, x3, x4 = powers(3)
print(x3)
27
```

Unnamed Functions (Lambda Function)

In Python we can also create unnamed functions, using the lambda keyword:

```
f1 = lambda x: x**2
# is equivalent to
def f2(x):
    return x**2
f1(2), f2(2)
(4, 4)
```

This technique is useful for example when we want to pass a simple function as an argument to another function, like this:

```
# map is a built-in python function
map(lambda x: x**2, range(-3,4))
[9, 4, 1, 0, 1, 4, 9]
# in python 3 we can use `list(...)` to convert the iterator to an explicit list
list(map(lambda x: x**2, range(-3,4)))
[9, 4, 1, 0, 1, 4, 9]
```

Classes

Classes are the key features of object-oriented programming. A class is a structure for representing an object and the operations that can be performed on the object.

In Python a class can contain *attributes* (variables) and *methods* (functions).

A class is defined almost like a function, but using the class keyword, and the class definition usually contains a number of class method definitions (a function in a class).

- Each class method should have an argument self as its first argument. This object is a self-reference.
- Some class method names have special meaning, for example:
  - *__init__*: The name of the method that is invoked when the object is first created.
  - *__str__* : A method that is invoked when a simple string representation of the class is needed, as for example when printed.
  - There are many more, see http://docs.python.org/2/reference/datamodel.html#special-method-names

```
class Point:
  """
  Simple class for representing a point in a Cartesian coordinate system.
  """
def __init__(self, x, y):
   """
```

```
        Create a new Point at x, y.
        """
        self.x = x
        self.y = y

    def translate(self, dx, dy):
        """
        Translate the point by dx and dy in the x and y direction.
        """
        self.x += dx
        self.y += dy

    def __str__(self):
        return("Point at [%f, %f]" % (self.x, self.y))
```

To create a new instance of a class:

```
p1 = Point(0, 0) # this will invoke the __init__ method in the Point class
print(p1)        # this will invoke the __str__ method
Point at [0.000000, 0.000000]
```

To invoke a class method in the class instance **p**:

```
p2 = Point(1, 1)
p1.translate(0.25, 1.5)
print(p1)
print(p2)
Point at [0.250000, 1.500000]
Point at [1.000000, 1.000000]
```

Note that calling class methods can modifiy the state of that particular class instance, but does not effect other class instances or any global variables. That is one of the nice things about object-oriented design: code such as functions and related variables are grouped in separate and independent entities.

Exceptions

In Python errors are managed with a special language construct called "Exceptions". When errors occur exceptions can be raised, which interrupts the normal program flow and fallback to somewhere else in the code where the closest try-except statement is defined. To generate an exception we can use the raise statement, which takes an argument that must be an instance of the class *BaseException* or a class derived from it.

```
raise Exception("description of the error")
```

```
---------------------------------------------------------------------------
Exception                              Traceback (most recent call last)
<ipython-input-128-8f47ba831d5a> in <module>()
----> 1 raise Exception("description of the error")
    Exception: description of the error
```

A typical use of exceptions is to abort functions when some error condition occurs, for example:

```
def my_function(arguments):
    if not verify(arguments):
        raise Exception("Invalid arguments")
        # rest of the code goes here
```

To gracefully catch errors that are generated by functions and class methods, or by the Python interpreter itself, use the try and except statements:

```
try:
    # normal code goes here
except:
    # code for error handling goes here
    # this code is not executed unless the code
    # above generated an error
```

For example:

```
try:
    print("test")
```

```
    # generate an error: the variable test is not defined
    print(test)
except:
    print("Caught an exception")
test
Caught an exception
```

To get information about the error, we can access the Exception class instance that describes the exception by using for example:

```
except Exception as e:
try:
    print("test")
    # generate an error: the variable test is not defined
    print(test)
except Exception as e:
    print("Caught an exception:" + str(e))
test
Caught an exception:name 'test' is not defined
```

NUMPY

We introduce the numpy module meant for numerical computation in this section. NumPy is a package that provide high-performance vector, matrix and higherdimensional data structures for Python. It is implemented in C and Fortran so when calculations are vectorized (formulated with vectors and matrices), performance is very good. To use numpy you need to import the module, using for example:

```
from numpy import *
```

In the numpy package the terminology used for vectors, matrices and higherdimensional data sets is *array*.

CREATING NUMPY ARRAYS

There are a number of ways to initialize new numpy arrays, for example from a Python list or tuples or using functions that are dedicated to generating numpy arrays (arange, linspace, etc.), or reading data from files. For example, to create new vector and matrix arrays from Python lists we can use the numpy.array function.

```
# a vector: the argument to the array function is a Python list
v = array([1,2,3,4])
v
array([1, 2, 3, 4])
# a matrix: the argument to the array function is a nested Python list
M = array([[1, 2], [3, 4]])
M
array([[1, 2], [3, 4]])
```

The v and M objects are both of the type ndarray that the numpy module provides.

```
type(v), type(M)
(numpy.ndarray, numpy.ndarray)
```

The difference between the v and M arrays is only their shapes. We can get information about the shape of an array by using the **ndarray.shape** property.

```
v.shape
(4,)
M.shape
(2, 2)
```

The number of elements in the array is available through the **ndarray.size** property:

```
M.size
4
```

Equivalently, we could use the functions **numpy.shape** and **numpy.size**

```
shape(M)
(2, 2)
size(M)
4
```

So far the numpy.ndarray looks awefully much like a Python list (or nested list). Why not simply use Python lists for computations instead of creating a new array type? There are several reasons:

- Python lists are very general. They can contain any kind of object. They are dynamically typed. They do not support mathematical functions such as matrix and dot multiplications, etc. Implementing such functions for Python lists would not be very efficient because of the dynamic typing.
- Numpy arrays are **statically typed** and **homogeneous**. The type of the elements is determined when the array is created.
- Numpy arrays are memory efficient.

Because of the static typing, fast implementation of mathematical functions such as multiplication and addition of numpy arrays can be implemented in a compiled language (C and Fortran is used).

Using the dtype (data type) property of an ndarray, we can see what type the data of an array has:

```
M.dtype
dtype('int64')
```

We get an error if we try to assign a value of the wrong type to an element in a numpy array:

```
M[0,0] = "hello"
---------------------------------------------------------------------------
ValueError Traceback (most recent call last)
<ipython-input-12-a09d72434238> in <module>()
----> 1 M[0,0] = "hello"
ValueError: invalid literal for long() with base 10: 'hello'
```

If we want, we can explicitly define the type of the array data when we create it, using the dtype keyword argument:

```
M = array([[1, 2], [3, 4]], dtype=complex)
```

```
M
array([[ 1.+0.j, 2.+0.j], [ 3.+0.j, 4.+0.j]])
```

Common data types that can be used with dtype are: int, float, complex, bool, object, etc. We can also explicitly define the bit size of the data types, for example: int64, int16, float128, complex128.

Using Array-Generating Functions

For larger arrays it is inpractical to initialize the data manually, using explicit python lists. Instead we can use one of the many functions in numpy that generate arrays of different forms. Some of the more common are:

arange

```
# create a range
x = arange(0, 10, 1) # arguments: start, stop, step
x
array([0, 1, 2, 3, 4, 5, 6, 7, 8, 9])
x = arange(-1, 1, 0.1)
x
array([ -1.00000000e+00, -9.00000000e-01, -8.00000000e-01,
-7.00000000e-01, -6.00000000e-01, -5.00000000e-01,
-4.00000000e-01, -3.00000000e-01, -2.00000000e-01,
-1.00000000e-01, -2.22044605e-16, 1.00000000e-01,
2.00000000e-01, 3.00000000e-01, 4.00000000e-01,
5.00000000e-01, 6.00000000e-01, 7.00000000e-01,
8.00000000e-01, 9.00000000e-01])
```

linspace and logspace

```
# using linspace, both end points ARE included
linspace(0, 10, 25)
array([ 0. , 0.41666667, 0.83333333, 1.25 ,
1.66666667, 2.08333333, 2.5 , 2.91666667,
3.33333333, 3.75 , 4.16666667, 4.58333333,
```

```
5. , 5.41666667, 5.83333333, 6.25 ,
6.66666667, 7.08333333, 7.5 , 7.91666667,
8.33333333, 8.75 , 9.16666667, 9.58333333, 10. ])
logspace(0, 10, 10, base=e)
array([ 1.00000000e+00, 3.03773178e+00, 9.22781435e+00,
2.80316249e+01, 8.51525577e+01, 2.58670631e+02,
7.85771994e+02, 2.38696456e+03, 7.25095809e+03,
2.20264658e+04])
mgrid
x, y = mgrid[0:5, 0:5] # similar to meshgrid in MATLAB
x
array([[0, 0, 0, 0, 0],
[1, 1, 1, 1, 1],
[2, 2, 2, 2, 2],
[3, 3, 3, 3, 3],
[4, 4, 4, 4, 4]])
y
array([[0, 1, 2, 3, 4],
[0, 1, 2, 3, 4],
[0, 1, 2, 3, 4],
[0, 1, 2, 3, 4],
[0, 1, 2, 3, 4]])
random data
from numpy import random
# uniform random numbers in [0,1]
random.rand(5,5)
array([[ 0.92932506, 0.19684255, 0.736434 , 0.18125714, 0.70905038],
[ 0.18803573, 0.9312815 , 0.1284532 , 0.38138008, 0.36646481],
[ 0.53700462, 0.02361381, 0.97760688, 0.73296701, 0.23042324],
[ 0.9024635 , 0.20860922, 0.67729644, 0.68386687, 0.49385729],
```

```
[ 0.95876515, 0.29341553, 0.37520629, 0.29194432, 0.64102804]])
# standard normal distributed random numbers
random.randn(5,5)
array([[ 0.117907 , -1.57016164, 0.78256246, 1.45386709, 0.54744436],
[2.30356897, -0.28352021, -0.9087325, 1.2285279, -1.00760167],
[0.72216801, 0.77507299, -0.37793178, -0.31852241, 0.84493629],
[-0.10682252, 1.15930142, -0.47291444, -0.69496967, -0.58912034],
[0.34513487, -0.92389516, -0.216978, 0.42153272, 0.86650101]])
diag
# a diagonal matrix
diag([1,2,3])
array([[1, 0, 0],
[0, 2, 0],
[0, 0, 3]])
# diagonal with offset from the main diagonal
diag([1,2,3], k=1)
array([[0, 1, 0, 0],
[0, 0, 2, 0],
[0, 0, 0, 3],
[0, 0, 0, 0]])
```

This short introduction should inspire the reader to delve deep into using NumPy. Many books are available and a lot of good tutorials are available on the web. Let us now have a brief look at the library of scientific algorithms in Python.

REFERENCES

1. Bioinformatics Programming - Using Python by Mitchell L. Model (O'Reilly, 2010)
2. Python Programming for Biology by Tim J. Stevens and Wayne Boucher (Cambridge University Press, 2015).

4

R IN BIOINFORMATICS

This introductory chapter is intended for you to get familiar with the statistical package R. R (www.r-project.org) is a commonly used open source statistical software with wide following and regular updates. **R** allows you to carry out statistical analyses in an interactive mode, as well as allowing simple to complex programming. We will learn the computational basics in this chapter and see the applications in the subsequent chapters. Let us start with installing the software on your computers.

INSTALLING R

To use R, you first need to install the R program on your computer.

The first thing to do is to check whether R is already installed on your computer (for example, by a previous user).

If you are using a Windows PC, there are two ways you can check whether R is already installed on your computer:

- Check if there is an "R" icon on the desktop of the computer that you are using. If so, double-click on the "R" icon to start R. If you cannot find an "R" icon, try step 2 instead.

- Click on the "Start" menu at the bottom left of your Windows desktop, and then move your mouse over "All Programs" in the menu that pops up. See if "R" appears in the list of programs that pops up. If it does, it means that R is already installed on your computer, and you can start R by selecting "R" (or *R X.X.X*, where *X.X.X* gives the version of R, e.g., *R 3.3.1*) from the list.

If either (1) or (2) above does succeed in starting R, it means that R is already installed on the computer that you are using. (If neither succeeds, R is not installed yet). If there is an old version of R installed on the Windows PC that you are using, it is worth installing the latest version of R, to make sure that you have all the latest R functions available to you to use.

These instructions will focus on installing R on a Windows PC. However, I will also briefly mention how to install R on a Macintosh or Linux computer (see below).

FINDING OUT WHAT IS THE LATEST VERSION OF R

To find out what is the latest version of R, you can look at the CRAN (Comprehensive R Network) website

https://cran.r-project.org/

You can find the latest release besides "The latest release" (about half way down the page). It will show something like "R-X.X.X.tar.gz" (e.g., "R-3.3.1.tar.gz"). This means that the latest release of R is X.X.X (for example, 3.3.1).

New releases of R are made very regularly (approximately once a month), as R is actively being improved all the time. It is worthwhile installing new versions of R regularly. Working the latest version of R is recommended for short-term projects (to ensure compatibility with all the latest versions of the R packages that you have downloaded).

A caution for those involved in long-term investigations. Try to maintain version consistency. Hence, complete your projects using a single version of **R**. Upon switching to a newer version of R, please make sure that the results are *reproducible* in its entirety.

INSTALLING R ON A WINDOWS PC

To install R on your Windows computer, follow these steps:

1. Go to *http://ftp.heanet.ie/mirrors/cran.r-project.org*
2. Under "Download and Install R", click on the "Windows" link.
3. Under "Subdirectories", click on the "base" link.
4. On the next page, you should see a link with something like "Download R 3.3.1 for Windows" (or R X.X.X, where X.X.X gives the version of R, e.g., R 3.3.1). Click on this link.
5. You may be asked if you want to save or run a file "R-3.3.1-win64.exe." Choose "Save" and save the file on the Desktop. Then double-click on the icon for the file to run it.
6. You will be asked what language to install it in - choose English.
7. The R Setup Wizard will appear in a window. Click "Next" at the bottom of the R Setup wizard window.
8. The next page says "Information" at the top. Click "Next" again.
9. The next page says "Information" at the top. Click "Next" again.
10. The next page says "Select Destination Location" at the top. By default, it will suggest to install R in "C:\Program Files" on your computer.
11. Click "Next" at the bottom of the R Setup wizard window.
12. The next page says "Select components" at the top. Click "Next" again.
13. The next page says "Startup options" at the top. Click "Next" again.
14. The next page says "Select start menu folder" at the top. Click "Next" again.

15. The next page says “Select additional tasks” at the top. Click “Next” again.

16. R should now be installed. This will take about a minute. When R has finished, you will see “Completing the R for Windows Setup Wizard” appear. Click “Finish”.

17. To start R, you can either follow step 18, or 19:

18. Check if there is an “R” icon on the desktop of the computer that you are using. If so, double-click on the “R” icon to start R. If you cannot find an “R” icon, try step 19 instead.

19. Click on the “Start” button at the bottom left of your computer screen, and then choose “All programs”, and start R by selecting “R” (or R X.X.X, where X.X.X gives the version of R, e.g., R 3.3.1) from the menu of programs.

20. The R console (a rectangle) should pop up:

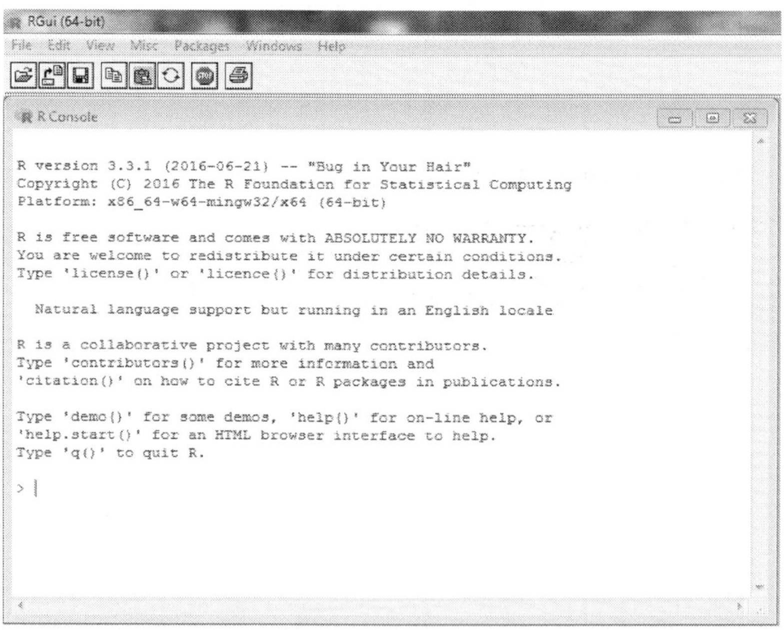

HOW TO INSTALL R ON NON-WINDOWS COMPUTERS

The instructions above are for installing R on a Windows PC. If you want to install R on a computer that has a non-Windows operating system (Macintosh or Linux), you should download the appropriate R installer for that operating system at

http://ftp.heanet.ie/mirrors/cran.r-project.org

and follow the R installation instructions for the appropriate operating system at

http://ftp.heanet.ie/mirrors/cran.r-project.org/doc/FAQ/R-FAQ.html#How-can-R-be-installed_003f

INSTALLING R PACKAGES

R comes with some standard packages that are installed when you install **R**. In this chapter we will also tell you how to use **R** packages that are useful in biomolecular science, for example, the "*rmeta*" package. These additional packages do not come with the standard installation of **R**. Hence, you need to install them by yourself.

HOW TO INSTALL AN R PACKAGE

Once you have installed **R** on a computer, you can install an additional package by following the steps below:

1. To start **R**, follow either step 2 or 3:
2. Check if there is an "**R**" icon on the desktop of the computer that you are using. If so, double-click on the "**R**" icon to start **R**. If you cannot find an "**R**" icon, try step 3 instead.
3. Click on the "Start" button at the bottom left of your computer screen, and then choose "All programs", and start **R** by selecting "**R**" (or *R X.X.X*, where *X.X.X* gives the version of **R**, e.g., *R 3.3.1*) from the menu of programs.
4. The **R** console (a rectangle) should pop up.

5. Once you have started **R**, you can now install an **R** package (e.g., the "*rmeta*" package) by choosing "Install package(s)" from the "Packages" menu at the top of the **R** console. This will ask you what website you want to download the package from, you should choose "USA" (or another country, if you prefer). It will also bring up a list of available packages that you can install, and you should choose the package that you want to install from that list (e.g., "*rmeta*").
6. This will install the "*rmeta*" package.
7. The "*rmeta*" package is now installed. Whenever you want to use the "*rmeta*" package after this, after starting **R**, you first have to load the package by typing into the **R** console:

```
> library("rmeta")
```

Note that there are some additional **R** packages for bioinformatics that are part of a special set of **R** packages called **Bioconductor** (www.bioconductor.org) such as the "*yeastExpData*" **R** package, the "*Biostrings*" **R** package, etc.). These **Bioconductor** packages need to be installed using a different, **Bioconductor**-specific procedure (see How to install a **Bioconductor R** package below).

HOW TO INSTALL A BIOCONDUCTOR R PACKAGE

The procedure above can be used to install the majority of **R** packages. However, the **Bioconductor** set of bioinformatics **R** packages need to be installed by a special procedure. **Bioconductor** (*www.bioconductor.org*) is a group of **R** packages that have been developed for bioinformatics. This includes R packages such as "*yeastExpData*", "*Biostrings*", etc.

To install the **Bioconductor** packages, follow these steps:

- To start R, follow either step 2 or 3:
- Check if there is an "**R**" icon on the desktop of the computer that you are using. If so, double-click on the "**R**" icon to start **R**. If you cannot find an "**R**" icon, try step 3 instead.
- Click on the "Start" button at the bottom left of your computer screen, and then choose "All programs", and start **R** by

selecting "**R**" (or *R X.X.X*, where *X.X.X* gives the version of **R**, e.g., *R 3.3.1*) from the menu of programs.

- The **R** console (a rectangle) should pop up.
- Once you have started **R**, now type in the **R** console:

```
> source("http://bioconductor.org/biocLite.R")
> biocLite()
```

- This will install a core set of **Bioconductor** packages (*"affy", "affydata", "affyPLM", "annaffy", "annotate", "Biobase", "Biostrings", "DynDoc", "gcrma", "genefilter", "geneplotter", "hgu95av2.db", "limma", "marray", "matchprobes", "multtest", "ROC", "vsn", "xtable", "affyQCReport"*). This takes a few minutes.
- At a later date, you may wish to install some extra Bioconductor packages that do not belong to the core set of Bioconductor packages. For example, to install the Bioconductor package called "*yeastExpData*", start **R** and type in the **R** console:

```
> source("http://bioconductor.org/biocLite.R")
> biocLite("yeastExpData")
```

- Whenever you want to use a package after installing it, you need to load it into **R** by typing:

```
> library("yeastExpData")
```

RUNNING R

To start **R**, you can either follow step 1 or 2:

1. Check if there is an "**R**" icon on the desktop of the computer that you are using. If so, double-click on the "**R**" icon to start **R**. If you cannot find an "**R**" icon, try step 2 instead.
2. Click on the "Start" button at the bottom left of your computer screen, and then choose "All programs", and start **R** by selecting "**R**" (or *R X.X.X*, where *X.X.X* gives the version of **R**, e.g., *R 3.3.1*) from the menu of programs.

This should bring up a new window, which is the *R console*.

RSTUDIO

To readers who are used to programming in Integrated Development Environment (IDE), *RStudio* is the best option. *RStudio* is a free open-source IDE for **R**. It was founded by J.J. Allaire and Hadley Wickham is the Chief Scientist at *RStudio*. It is available in two versions.

- *RStudio Desktop*: program is run locally on a desktop computer or laptop.
- *RStudio Server*: allows accessing *RStudio* using a web browser while it is running on a remote Linux server.

It is also available as a commercial edition. Please follow the installation instructions at:

www.rstudio.com

This is an alternative to working on **R** console. RStudio requires prior installation of **R**.

We start with a brief introduction to R, with hands-on training of basic mathematical and statistical operations in R.

A BRIEF INTRODUCTION TO R

You will type **R** commands into the **R** console in order to carry out analyses in **R**. In the **R** console you will see:

```
>
```

This is the **R** prompt. We type the commands needed for a particular task after this prompt. The command is carried out after you hit the Return key. Once you have started **R**, you can start typing in commands, and the results will be calculated immediately. Try the following.

```
> 3*3
[1] 9
> 10-4
[1] 6
```

All variables (scalars, vectors, matrices, etc.) created by **R** are called *objects*. In **R**, we assign values to variables using an arrow. For example, we can assign the value 3*3 to the variable *x* using the command:

> x <- 3*3

To view the contents of any **R** object, just type its name, and the contents of that **R** object will be displayed:

> x

[1] 6

There are several possible different types of objects in **R**, including *scalars*, *vectors*, *matrices*, *arrays*, *dataframes*, *tables*, and *lists*. The scalar variable *x* above is one example of an **R** object. While a scalar variable such as *x* has just one element, a vector consists of several elements. The elements in a vector are all of the same type (e.g., *numeric* or *characters*), while lists may include elements such as characters as well as numeric quantities.

To create a vector, we can use the *c()* (combine) function. For example, to create a vector called *vector_1* that has elements with values 10, 6, 9 and 5, we type:

> vector_1 <- c(10, 6, 9, 5)

To see the contents of the variable *vector_1*, we can just type its name:

> vector_1

[1] 10 6 9 5

The [1] is the index of the first element in the vector. We can extract any element of the vector by typing the vector name with the index of that element given in square brackets. For example, to get the value of the 2nd element in the vector *vector_1*, we type:

> vector_1[2]

[1] 6

In contrast to a *vector*, a *list* can contain elements of different types, for example, both numeric and character elements. A list can also include other variables such as a vector. The *list()* function is used to create a list. Let us create a list *list* _1 as follows:

> list_1 <- list(name="Deepak", wife="Shalini", vector_1)

We can then print out the contents of the list *list_1* by typing its name:

> list_1

$name

[1] "Deepak"

$wife

[1] "Shalini"

[[3]]

[1] 10 6 9 5

The elements in a list are numbered, and can be referred to using indices. We can extract an element of a list by typing the list name with the index of the element given in double square brackets (in contrast to a vector, where we only use single square brackets). Thus, we can extract the second and third elements from *list _1* by typing:

> list_1[[2]]

[1] "Shalini"

> list_1[[3]]

[1] 10 6 9 5

Elements of lists may also be named, and in this case the elements may be referred to by giving the list name, followed by "$", followed by the element name. For example, *list_1$name* is the same as *list_1[[1]]* and *list_1$wife* is the same as *list_1[[2]]*:

> list_1$wife

[1] "Shalini"

We can find out the names of the named elements in a list by using the *attributes ()* function, for example:

> attributes(list_1)

$names

[1] "name" "wife" ""

When you use the *attributes ()* function to find the named elements of a list variable, the named elements are always listed under a heading "$names." Therefore, we see that the named elements of the list variable *list_1* are called "name" and "wife", and we can retrieve their values by typing *list_1$name* and *list_1$wife*, respectively.

Another type of object that you will encounter in **R** is a *table* variable. For example, if we made a vector variable *mynames* containing the names of children in a class, we can use the *table ()* function to produce a table variable that contains the number of children with each possible name:

> mynames <- c("Mary", "John", "Ann", "Sinead", "Joe", "Mary", "Jim", "John", "Simon")

> table(mynames)

mynames

```
 Ann   Jim   Joe  John  Mary Simon Sinead
   1     1     1     2     2     1      1
```

We can store the table variable produced by the function *table ()*, and call the stored table "mytable", by typing:

> mytable <- table(mynames)

To access elements in a table variable, you need to use double square brackets, just like accessing elements in a list. For example, to access the fourth element in the table *mytable* (the number of children called "John"), we type:

> mytable[[4]]

[1] 2

Alternatively, you can use the name of the fourth element in the table ("John") to find the value of that table element:

> mytable[["John"]]

[1] 2

Functions in **R** usually require *arguments*, which are input variables (i.e., objects) that are passed to them, which they then carry out some operation on. For example, the *log10 ()* function is passed a number, and it then calculates the *log* to the *base 10* of that number:

> log10(100)

2

In **R**, you can get help about a particular function by using the *help ()* function. For example, if you want help about the *log10 ()* function, you can type:

> help("log10")

When you use the *help ()* function, a box or webpage will pop up with information about the function that you asked for help with.

If you are not sure of the name of a function, but think you know part of its name, you can search for the function name using the *help. Search ()* and *RSiteSearch()* functions. The *help.search()* function searches to see if you already have a function installed (from one of the **R** packages that you have installed) that may be related to some topic you're interested in. The *RSiteSearch()* function searches all **R** functions (including those in packages that you haven't yet installed) for functions related to the topic you are interested in.

For example, if you want to know if there is a function to calculate the standard deviation of a set of numbers, you can search for the names of all installed functions containing the word "deviation" in their description by typing:

> help.search("deviation")

Help files with alias or concept or title matching 'deviation' using fuzzy matching:

genefilter::rowSds

Row variance and standard deviation of a numeric array

nlme::pooledSD Extract Pooled Standard Deviation

stats::mad Median Absolute Deviation

stats::sd Standard Deviation

vsn::meanSdPlot Plot row standard deviations versus row

Among the functions that were found, is the function *sd()* in the "*stats*" package (an **R** package that comes with the standard **R** installation), which is used for calculating the standard deviation.

In the example above, the *help.search()* function found a relevant function (*sd()* here). However, if you did not find what you were looking for with *help.search()*, you could then use the *RSiteSearch()* function to see if a search of all functions described on the **R** website may find something relevant to the topic that you're interested in:

> RSiteSearch("deviation")

The results of the *RSiteSearch()* function will be hits to descriptions of **R** functions, as well as to **R** mailing list discussions of those functions.

We can perform computations with **R** using objects such as scalars and vectors. For example, to calculate the average of the values in the vector *myvector* (i.e., the average of 10, 6, 9 and 5), we can use the *mean()* function:

```
> mean(myvector)
[1] 7.5
```

We have been using built-in **R** functions such as *mean()*, *length()*, print(), *plot()* etc. We can also create our own functions in **R** to do calculations that you want to carry out very often on different input data sets. For example, we can create a function to calculate the value of 20 plus square of some input number:

```
> myfunction <- function(x) { return(20 + (x*x)) }
```

This function will calculate the square of a number (*x*), and then add 20 to that value. The *return()* statement returns the calculated value. Once you have typed in this function, the function is then available for use. For example, we can use the function for different input numbers (e.g., 10, 25):

```
> myfunction(10)
[1] 120
> myfunction(25)
[1] 645
```

To quit R, type:

```
> q()
```

BIOCONDUCTOR AND SEQINR

Many authors have written R packages for performing a wide variety of analyses. These do not come with the standard R installation, but must be installed and loaded as "*add-ons*." Bioinformaticians have written several specialised *packages* for R. In this section, you will learn to use the SeqinR package to retrieve sequences from a DNA sequence database, and to carry out simple analyses of DNA sequences. Some well-known bioinformatics packages for R are the Bioconductor set of R packages (*www.bioconductor.org*), which contains several packages with many R functions for analysing biological data sets such as microarray data; and the SeqinR package (*pbil.univ-lyon1.fr/software/seqinr/home.php?lang=eng*), which contains R functions for obtaining sequences from DNA and protein sequence databases, and for analysing DNA and protein sequences.

To use function from the SeqinR package, we first need to install the SeqinR package (for instructions on how to install an R package, see above). Once you have installed the "SeqinR" R package, you can load the "SeqinR" R package by typing:

> library("seqinr")

Remember that you can ask for more information about a particular **R** command by using the *help()* function. For example, to ask for more information about the *library()* function, you can type:

> help("library")

THE NCBI SEQUENCE DATABASE

We have already seen what NCBI stands for and biological sequence data is stored in it. We have also seen how to query the NCBI database for information. The National Centre for Biotechnology Information (NCBI) (*www.ncbi.nlm.nih.gov*) in the US maintains a huge database of all the DNA and protein sequence data that has been collected, the NCBI Sequence Database. This also a similar database in Europe, the European Molecular Biology Laboratory (EMBL) Sequence Database (*www.ebi.ac.uk/embl*), and also a similar database in Japan, the DNA Data Bank of Japan (DDBJ; *www.ddbj.nig.ac.jp*). These three databases exchange data every night, so at any one point in time, they contain almost identical data. Interested readers are requested to delve into these resources.

Let us get into the details of how sequences are stored in NCBI. Each sequence in the NCBI Sequence Database is stored in a separate *record*, and is assigned a unique identifier that can be used to refer to that sequence record. The identifier is known as an *accession*, and consists of a mixture of numbers and letters. For example, Dengue virus causes Dengue fever, which is classified as a neglected tropical disease by the WHO. by any one of four types of Dengue virus: DEN-1, DEN-2, DEN-3, and DEN-4. The NCBI accessions for the DNA sequences of the

DEN-1, DEN-2, DEN-3, and DEN-4 Dengue viruses are NC_001477, NC_001474, NC_001475 and NC_002640, respectively.

Note that because the NCBI Sequence Database, the EMBL Sequence Database, and DDBJ exchange data every night, the DEN-1 (and DEN-2, DEN-3, DEN-4) Dengue virus sequence will be present in all three databases, but it will have different accessions in each database, as they each use their own numbering systems for referring to their own sequence records.

RETRIEVING GENOME SEQUENCE DATA VIA THE NCBI WEBSITE

You can easily retrieve DNA or protein sequence data from the NCBI Sequence Database *via* its website.

The Dengue DEN-1 DNA sequence is a viral DNA sequence, and as mentioned above, its NCBI accession is NC_001477. To retrieve the DNA sequence for the Dengue DEN-1 virus from NCBI, go to the NCBI website, type "NC_001477" in the Search box at the top of the webpage, and press the "Search" button beside the Search box:

On the results page you will see the number of hits to "NC_001477" in each of the NCBI databases on the NCBI website. There are many databases on the NCBI website, for example, the "*PubMed*" data contains abstracts from scientific papers, the "*Nucleotide*" database contains DNA and RNA sequence data, the "*Protein*" data contains protein sequence data, and so on. The picture below shows what the results page should look like for your NC_001477 search. As you are looking for the DNA sequence of the Dengue DEN-1 virus genome, you expect to see a hit in the NCBI Nucleotide database, and indeed there is a hit in the Nucleotide database (indicated by the "1" beside the icon for the Nucleotide database):

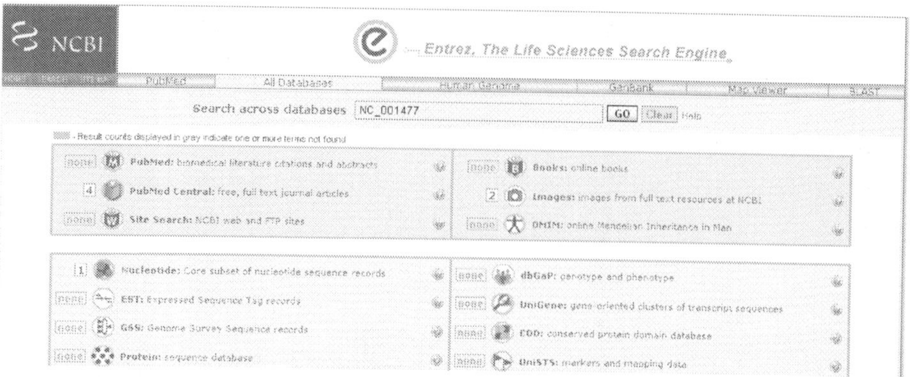

To look at the one sequence found in the Nucleotide database, you need to click on the icon for the NCBI Nucleotide database on the results page for the search:

When you click on the icon for the NCBI Nucleotide database, it will bring you to the record for NC_001477 in the NCBI Nucleotide database. This will contain the name and NCBI accession of the sequence, as well as other details such as any papers describing the sequence:

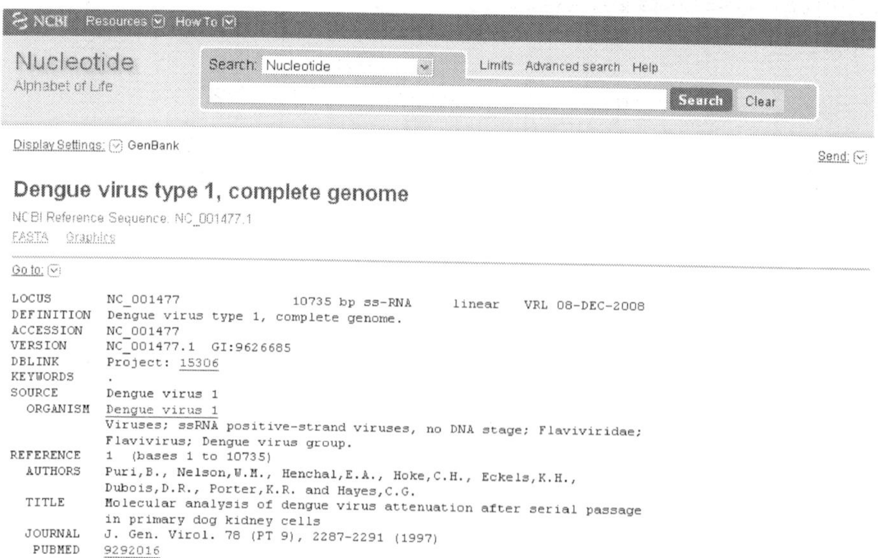

To retrieve the DNA sequence for the DEN-1 Dengue virus genome sequence as a FASTA format sequence file, click on "*Send*" at the top right of the NC_001477 sequence record webpage, and then choose "*File*" in the pop-up menu that appears, and then choose FASTA from the "Format" menu that appears, and click on "*Create file*".

A box will pop up asking you what to name the file, and where to save it. You should give it a sensible name (e.g., "den1.fasta") and save it in a place where you will remember (e.g., in the "My Documents" folder is a good idea):

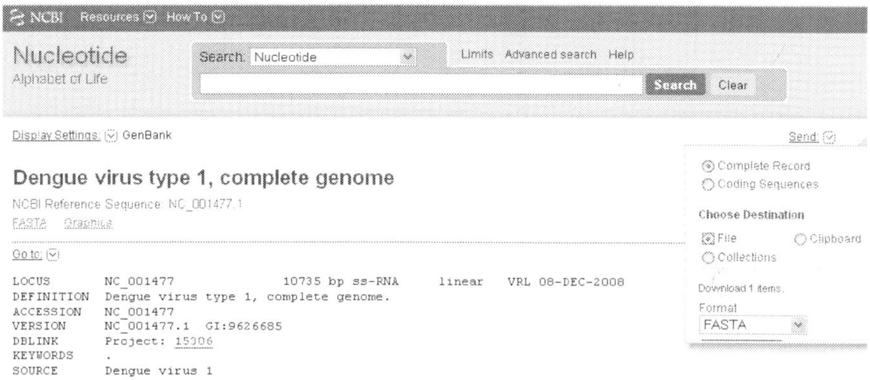

You can now open the FASTA file containing the DEN-1 Dengue virus genome sequence using WordPad on your computer. To open WordPad, click on "Start" on the bottom left of your screen, click on "All Programs" in the menu that appears, and then select "Accessories" from the menu that appears next, and then select "WordPad" from the menu that appears next. WordPad should start up. In Wordpad, choose "Open" from the "File" menu. The WordPad "Open" dialog will appear. Set "Files of type" to "All Documents" at the bottom of the WordPad "Open" dialog. You should see a list of files, now select the file that contains the DEN-1 Dengue virus sequence (e.g., "den1.fasta"). The contents of the FASTA format file containing the Dengue DEN-1 sequence should now be displayed in WordPad:

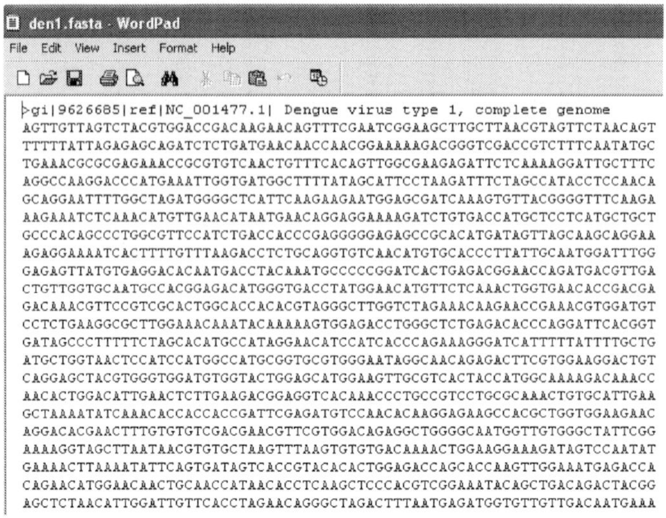


```
>gi|9626685|ref|NC_001477.1| Dengue virus type 1, complete genome
AGTTGTTAGTCTACGTGGACCGACAAGAACAGTTTCGAATCGGAAGCTTGCTTAACGTAGTTCTAACAGT
TTTTTATTAGAGAGCAGATCTCTGATGAACAACCAACGGAAAAAGACGGGTCGACCGTCTTTCAATATGC
TGAAACGCGCGAGAAACCGCGTGTCAACTGTTTCACAGTTGGCGAAGAGATTCTCAAAAGGATTGCTTTC
AGGCCAAGGACCCATGAAATTGGTGATGGCTTTTATAGCATTCCTAAGATTTCTAGCCATACCTCCAACA
GCAGGAATTTTGGCTAGATGGGGCTCATTCAAGAAGAATGGAGCGATCAAAGTGTTACGGGGTTTCAAGA
AAGAAATCTCAAACATGTTGAACATAATGAACAGGAGGAAAAGATCTGTGACCATGCTCCTCATGCTGCT
GCCCACAGCCCTGGCGTTCCATCTGACCACCCGAGGGGGAGAGCCGCACATGATAGTTAGCAAGCAGGAA
AGAGGAAAATCACTTTTGTTTAAGACCTCTGCAGGTGTCAACATGTGCACCCTTATTGCAATGGATTTGG
GAGAGTTATGTGAGGACACAATGACCTACAAATGCCCCCGGATCACTGAGACGGAACCAGATGACGTTGA
CTGTTGGTGCAATGCCACGGAGACATGGGTGACCTATGGAACATGTTCTCAAACTGGTGAACACCGACGA
GACAAACGTTCCGTCGCACTGGCACCACACGTAGGGCTTGGTCTAGAAACAAGAACCGAAACGTGGATGT
CCTCTGAAGGCGCTTGGAAACAAATACAAAAAGTGGAGACCTGGGCTCTGAGACACCCAGGATTCACGGT
GATAGCCCTTTTTCTAGCACATGCCATAGGAACATCCATCACCCAGAAAGGGATCATTTTTATTTTGCTG
ATGCTGGTAACTCCATCCATGGCCATGCGGTGCGTGGGAATAGGCAACAGAGACTTCGTGGAAGGACTGT
CAGGAGCTACGTGGGTGGATGTGGTACTGGAGCATGGAAGTTGCGTCACTACCATGGCAAAAGACAAACC
AACACTGGACATTGAACTCTTGAAGACGGAGGTCACAAACCCTGCCGTCCTGCGCAAACTGTGCATTGAA
GCTAAAATATCAAACACCACCACCGATTCGAGATGTCCAACACAAGGAGAAGCCACGCTGGTGGAAGAAC
AGGACACGAACTTTGTGTGTCGACGAACGTTCGTGGACAGAGGCTGGGGCAATGGTTGTGGGCTATTCGG
AAAAGGTAGCTTAATAACGTGTGCTAAGTTTAAGTGTGTGACAAAACTGGAAGGAAAGATAGTCCAATAT
GAAAACTTAAAATATTCAGTGATAGTCACCGTACACACTGGAGACCAGCACCAAGTTGGAAATGAGACCA
CAGAACATGGAACAACTGCAACCATAACACCTCAAGCTCCCACGTCGGAAATACAGCTGACAGACTACGG
AGCTCTAACATTGGATTGTTCACCTAGAACAGGGCTAGACTTTAATGAGATGGTGTTGTTGACAATGAAA
```

RETRIEVING GENOME SEQUENCE DATA USING SEQINR

Instead of going to the NCBI website to retrieve sequence data from the NCBI database, you can retrieve sequence data from NCBI directly from **R**, by using the **SeqinR R** package. For example, you learnt above how to retrieve the DEN-1 Dengue virus genome sequence, which has NCBI accession NC_001477, from the NCBI website. To retrieve a sequence with a particular NCBI accession, you can use **R** function "*getncbiseq()*" below, which you will first need to copy and paste into R:

```
> getncbiseq <- function(accession)
  {
    require("seqinr") # this function requires the SeqinR R package
    # first find which ACNUC database the accession is stored in:
    dbs <- c("genbank","refseq","refseqViruses","bacterial")
    numdbs <- length(dbs)
    for (i in 1:numdbs)
    {
      db <- dbs[i]
```

```
    choosebank(db)
    # check if the sequence is in ACNUC database 'db':
    resquery <- try(query(".tmpquery", paste("AC=", accession)), silent
= TRUE)
    if (!(inherits(resquery, "try-error")))
    {
      queryname <- "query2"
      thequery <- paste("AC=",accession,sep="")
      query(`queryname`,`thequery`)
      # see if a sequence was retrieved:
      seq <- getSequence(query2$req[[1]])
      closebank()
      return(seq)
    }
    closebank()
  }
  print(paste("ERROR: accession",accession,"was not found"))
}
```

Once you have copied and pasted the function *getncbiseq()* into **R**, you can use it to retrieve a sequence from the NCBI Nucleotide database, such as the sequence for the DEN-1 Dengue virus (accession NC_001477):

```
> dengueseq <- getncbiseq("NC_001477")
```

The variable *dengueseq* is a vector containing the nucleotide sequence. Each element of the vector contains one nucleotide of the sequence. Therefore, to print out a certain subsequence of the sequence, we just need to type the name of the vector *dengueseq* followed by the square brackets containing the indices for those nucleotides. For example, the following command prints out the first 50 nucleotides of the DEN-1 Dengue virus genome sequence:

```
> dengueseq[1:50]
```

[1] "a" "g" "t" "t" "g" "t" "t" "a" "g" "t" "c" "t" "a" "c" "g" "t" "g" "g" "a"

[20] "c" "c" "g" "a" "c" "a" "a" "g" "a" "a" "c" "a" "g" "t" "t" "t" "c" "g" "a"

[39] "a" "t" "c" "g" "g" "a" "a" "g" "c" "t" "t" "g"

Note that *dengueseq[1:50]* refers to the elements of the vector *dengueseq* with indices from 1-50. These elements contain the first 50 nucleotides of the DEN-1 Dengue virus sequence.

WRITING SEQUENCE DATA OUT AS A FASTA FILE

If you have retrieved a sequence from the NCBI database using the "*getncbiseq()*" function, you may want to save the sequence to a FASTA-format file on your computer, in case you need the sequence for further analyses (either in **R** or in other software).

You can write out a sequence to a FASTA-format file in **R** by using the "*write.fasta()*" function from the **SeqinR R** package. The *write.fasta()* function requires that you tell it the name of the output file using the "*file.out*" argument (input). You also need to specify the **R** variable that contains the sequence using the "*sequences*" argument, and the name that you want to give to the sequence using the "*names*" argument. For example, if you have stored the DEN-1 Dengue virus sequence in a vector *dengueseq*, to write out the sequence to a FASTA-format file called "den1.fasta" that contains the sequence labelled as "DEN-1", you can type:

> write.fasta(names="DEN-1", sequences=dengueseq, file.out="den1.fasta")

READING SEQUENCE DATA INTO R

Using the **SeqinR** package in **R**, you can easily read a DNA sequence from a FASTA file into **R**. For example, we described above how to retrieve the DEN-1 Dengue virus genome sequence from the NCBI database, or from **R** using the *getncbiseq()* function, and save it in a FASTA format file (e.g., "den1.fasta"). You can read this FASTA format file into **R** using the *read.fasta()* function from the **SeqinR R** package:

```
> library("seqinr")
> dengue <- read.fasta(file = "den1.fasta")
```

Note that **R** expects the files that you read in (e.g., "den1.fasta") to be in the "My Documents" folder on your computer, so if you stored "den1.fasta" somewhere else, you will have to move or copy it into "My Documents." The above command reads the contents of the fasta format file den1.fasta into an **R** object called *dengue*. The variable *dengue* is an **R** list object. As explained above, a *list* is an R object that is like a vector, but can contain elements that are numeric and/or contain characters. In this case, the list *dengue* contains information from the FASTA file that you have read in (i.e., the name given to the dengue sequence in the FASTA file, and the DNA sequence itself). In fact, the first element of the list object *dengue* contains the DNA sequence. We can access the elements of an **R** list object using double square brackets. Thus, we can store the DNA sequence for DEN-1 Dengue virus in a variable *dengueseq* by typing:

```
> dengueseq <- dengue[[1]]
```

Now the variable *dengueseq* is a vector containing the nucleotide sequence.

Length of a DNA sequence

Once you have retrieved a DNA sequence, we can obtain some simple statistics to describe that sequence, such as the sequence's total length in nucleotides. In the above example, we retrieved the DEN-1 Dengue virus genome sequence, and stored it in the vector variable *dengueseq*. To subsequently obtain the length of the genome sequence, we would use the *length()* function, typing:

```
> length(dengueseq)
[1] 10735
```

The *length()* function will give you back the length of the sequence stored in variable *dengueseq*, in nucleotides. The *length()* function actually gives the number of elements in the input vector that you pass to it, which in this case in the number of elements in the vector *dengueseq*. Since each element of the vector *dengueseq* contains one nucleotide of the DEN-1 Dengue virus sequence, the result for the DEN-1 Dengue virus genome tells us the length of its genome sequence (i.e., 10735 nucleotides long).

Base composition of a DNA sequence

An obvious first analysis of any DNA sequence is to count the number of occurrences of the four different nucleotides ("A", "C", "G", and "T") in the sequence. This can be done using the *table()* function. For example, to find the number of As, Cs, Gs, and Ts in the DEN-1 Dengue virus sequence (which you have put into vector variable *dengueseq*, using the commands above), you would type:

```
> table(dengueseq)
dengueseq
   a    c    g    t
3426 2240 2770 2299
```

This means that the DEN-1 Dengue virus genome sequence has 3426 As, 2240 Cs, 2770 Gs and 2299 Ts.

GC Content of DNA

One of the most fundamental properties of a genome sequence is its GC content, the fraction of the sequence that consists of Gs and Cs, i.e., the %(G+C). The GC content can be calculated as the percentage of the bases in the genome that are Gs or Cs. That is,

GC content = (number of Gs + number of Cs)*100/(genome length)

For example, if the genome is 100 bp, and 20 bases are Gs and 21 bases are Cs, then the GC content is (20 + 21)*100/100 = 41%.

You can easily calculate the GC content based on the number of As, Gs, Cs, and Ts in the genome sequence. For example, for the DEN-1 Dengue virus genome sequence, we know from using the table() function above that the genome contains 3426 As, 2240 Cs, 2770 Gs and 2299 Ts. Therefore, we can calculate the GC content using the command:

```
> (2240+2770)*100/(3426+2240+2770+2299)
[1] 46.66977
```

Alternatively, if you are feeling lazy, you can use the *GC()* function in the SeqinR package, which gives the fraction of bases in the sequence that are Gs or Cs.

```
> GC(dengueseq)
[1] 0.4666977
```

The result above means that the fraction of bases in the DEN-1 Dengue virus genome that are Gs or Cs is 0.4666977. To convert the fraction to a percentage, we have to multiply by 100, so the GC content as a percentage is 46.66977%.

DNA words

As well as the frequency of each of the individual nucleotides ("A", "G", "T", "C") in a DNA sequence, it is also interesting to know the frequency of longer DNA "words." The individual nucleotides are DNA words that are 1 nucleotide long, but we may also want to find out the frequency of DNA words that are 2 nucleotides long (i.e., "AA", "AG", "AC", "AT", "CA", "CG", "CC", "CT", "GA", "GG", "GC", "GT", "TA", "TG", "TC", and "TT"), 3 nucleotides long (e.g., "AAA", "AAT", "ACG", etc.), 4 nucleotides long, etc.

To find the number of occurrences of DNA words of a particular length, we can use the count() function from the R SeqinR package. For example, to find the number of occurrences of DNA words that are 1 nucleotide long in the sequence *dengueseq*, we type:

```
> count(dengueseq, 1)
 a c g t
3426 2240 2770 2299
```

As expected, this gives us the number of occurrences of the individual nucleotides. To find the number of occurrences of DNA words that are 2 nucleotides long, we type:

```
> count(dengueseq, 2)
 aa ac ag at ca cc cg ct ga gc gg gt ta tc tg tt
 1108 720 890 708 901 523 261 555 976 500 787 507 440 497
832 529
```

Note that by default the count() function includes all overlapping DNA words in a sequence. Therefore, for example, the sequence "ATG" is considered to contain two words that are two nucleotides long: "AT" and "TG".

If you type help('count'), you will see that the result (output) of the function count() is a *table* object. This means that you can use double square brackets to extract the values of elements from the table. For example, to extract the value of the third element (the number of Gs in the DEN-1 Dengue virus sequence), you can type:

```
> denguetable <- count(dengueseq,1)
> denguetable[[3]]
 [1] 2770
```

The command above extracts the third element of the table produced by count(dengueseq,1), which we have stored in the table variable *denguetable*.

Alternatively, you can find the value of the element of the table in column "g" by typing:

```
> denguetable[["g"]]
 [1] 2770
```

SUMMARY

In the previous sections, you have learnt to use the following R functions:

- *length()* for finding the length of a vector or list
- *table()* for printing out a table of the number of occurrences of each type of item in a vector or list.

These functions belong to the standard installation of **R**.

You have also learnt the following **R** functions that belong to the **SeqinR** package:

- *GC()* for calculating the GC content for a DNA sequence
- *count()* for calculating the number of occurrences of DNA words of a particular length in a DNA sequence

REFERENCES

- *Little Book of R for Bioinformatics* by Avril Coghlan, Wellcome Trust Sanger Institute, Cambridge, U.K (**2016**).
- Charif D & Lobry JR. "SeqinR 1.0-2: a contributed package to the R project for Statistical Computing Devoted to Biological Sequences Retrieval and Analysis in Structural approaches to sequence evolution: Molecules, networks, populations (**2007**).

5

SEQUENCE ANALYSIS

INTRODUCTION

Biological sequence analysis long predates the Human Genome Project (HGP). However, it has recently come into its own with the exponential growth in sequence information generated by the HGP. The idea of undertaking a coordinated international study of the entire human genetic complement arose from a series of meetings held between 1985 and 1987. The goal of the genome project, an informal worldwide scientific collaboration, is the fundamental characterization of the human genetic blueprint. Specifically, the HGP effort has been divided into two main parts:

- *Mapping*: in which biologically important regions of the genome (especially genes) are localized on the chromosomes relative to a framework of landmarks; and
- *Sequencing*: in which the genetic information encoded by the order of the DNA's chemical subunits is determined.

The ultimate goal of genomic mapping and sequencing is to associate specific human traits and diseases with one or more genes at precise locations on the chromosomes. The successful completion of the genome project will provide an unparalleled understanding of the fundamental organization of human genes and chromosomes. Consequently, it promises to revolutionize both **therapeutic** and **preventive medicine** by providing insights into the basic *biochemical processes* that underlie human disease.

This chapter provides a general introduction to biological sequence analysis. In order to keep the discussion brief, many topics have been omitted, and others have been only touched upon. The goal is to provide the reader with enough background to understand the biological underpinnings of the chapters that follow, along with some notion of the value of the analyses undertaken.

First, a little bit about the approach taken in this chapter and the ones that follow. We will be using computer language extensively for discussing and illustrating the concepts. It is assumed that the readers will get familiarized with these in order to understand what follows. A small introduction into the fundamentals has been given in the two previous chapters. We will be extensively using Python in this and following chapters.

Python is by all means a beautiful and simple computer language. In recent years, its popularity in bioinformatics community has increased substantially. Technically, Python is a powerful language with respect to computing paradigms. It has the skeleton of a *procedural* languages (like Prolog) with significant influence from *functional* languages (like LISP) and is a superb object-oriented language (like C++). The reader is encouraged to get started in Python using popular texts in the area. I personally like the following, for its organization and lucid style of writing:

Bioinformatics: Sequence and Genome Analysis, by David W. Mount (Cold Spring Harbor Laboratory Press)

- *Bioinformatics and Functional Genomics*, by Jonathan Pevsner (Wiley-Blackwell)
- *Learning Python* - Powerful Object-Oriented Programming, by Mark Lutz (O'Reilly Media)

We now take first steps in comprehending and performing basic analysis on biological sequences. We take an approach, wherein the reader gets to perform some short computing tutorials before delving into the technical details of methods. We start with a short introduction to working with sequences using

BioPython. *BioPython* is a set of libraries to provide the ability to deal with "things" of interest to biologists working on the computer. In general this means that you will need to have at least some programming experience or at least an interest in learning to program. Biopython's job is to make your job easier as a programmer by supplying reusable libraries so that you can *focus* on answering your *specific question* of interest, instead of focusing on the internals of *parsing* a particular file format. One thing to note about *BioPython* is that it often provides multiple ways of "doing the same thing." Things have improved in recent releases, but this can still be frustrating as in Python there should ideally be one right way to do something. However, this can also be a real benefit because it gives you lots of flexibility and control over the libraries. The tutorial helps to show you the common or easy ways to do things so that you can just make things work.

A BIT OF MOLEUCLAR BIOLOGY

In biological research, by sequences we mean polymeric substances. Proteins and Nucleic acids belong to this class of biomolecules. The fundamental processes of molecular biology, for our discussion, can be summarised below:

- **Deoxyribose Nucleic Acid (DNA):** The human genetic code is captured in long molecules of **DNA,** which are packaged in the cell nucleus in the chromosomes. **DNA** is composed of a long chain of subunits; the nucleotides adenine, guanine, cytosine, and thymine (abbreviated **A, G, C** and **T,** respectively). **DNA** is usually found in the form of two paired chains, held together by hydrogen bonds between complementary pairs of nucleotides. **A** on one strand always pairs with **T** on the other, and similarly for **G** and **C**.
- **Proteins**: Most of the important structural and functional components of a human cell are composed of proteins, which are long, folded chains composed of the 20 common

amino acids, or peptides. (Thus, the frequent use of the term "**polypeptide**" for a protein.)

- **Genes**: Genes are individual stretches of the chromosomal **DNA** that encode the sequence of amino acids comprising a particular protein. The **64** possible nucleotide triplets, known as **codons**, are used to represent the 20 amino acids using a degenerate code. In higher organisms, portions of genes that code for proteins ("**exons**") are usually separated by noncoding regions ("**introns**").
- **Transcription & Translation**: An intermediary form of nucleic acid called **RNA *transcribes*** a gene's **DNA** sequence in a form which can be directly ***translated*** by the cell's protein-making machinery (*ribosome*). Prior to processing an **RNA** sequence, the intronic regions are spliced out to leave just the coding portions to be translated.

The basic molecular structures of the above-mentioned biomolecules have been introduced in the previous chapter. Now, we move to perform some basic sequence based operations using **BioPython**.

```
from Bio.Seq import Seq
my_seq = Seq("AGTACACTGGT") # we have not specified whether it is a DNA or Protein sequence
my_seq
Seq('AGTACACTGGT', Alphabet())
print(my_seq)
AGTACACTGGT
my_seq.alphabet
Alphabet()
```

What we have here is a sequence object with a *generic* alphabet - reflecting the fact we have not specified if this is a DNA or protein sequence.

In addition to having an alphabet, the Seq object differs from the Python string in the methods it supports. You can't do this with a plain string:

```
my_seq.complement()
```

```
Seq('TCATGTGACCA', Alphabet())
my_seq.reverse_complement()
Seq('ACCAGTGTACT', Alphabet())
```

The next most important class is the **SeqRecord** or Sequence Record. This holds a sequence (as a **Seq** object) with additional annotation including an identifier, name and description. The **Bio.SeqIO** module for reading and writing sequence file formats works with SeqRecord objects, which will be introduced below and covered in more detail later. This covers the basic features and uses of the Biopython sequence class. Now that you've got some idea of what it is like to interact with the Biopython libraries, it's time to delve into biological file formats!

Let's start by taking a simple example from *Arabidopsis thaliana*. *A. thaliana* is a small flowering plant native to Eurasia. *A. thaliana* is considered a weed; it is found by roadsides and in disturbed land. A winter annual with a relatively short life cycle, *A. thaliana* is a popular **model organism** in plant biology and genetics. For a complex multicellular eukaryote, *A. thaliana* has a relatively small genome of approximately 135 megabase pairs (Mbp). It was the first plant to have its genome sequenced, and is a popular tool for understanding the molecular biology of many plant traits, including flower development and light sensing.

Arabidopsis thaliana (A. thaliana) is a plant discovered in the sixteenth century by Johannes Thal in the Harz Mountains (Germany). It is a member of the mustard (*Brassicaceae* or *Cruciferae*) family of dicotyledonous plants, which includes species such as cabbage and radish. *A. thaliana* is an annual herbaceous plant native to Europe, Central Asia and Northwest Africa. *Arabidopsis thaliana* belongs to the *Cruciferae* (family *Brassicaceae*, Capparales). *A. thaliana* has a rapid life cycle corresponding to approximately 6 weeks from germination to maturity.

Arabidopsis thaliana was the first plant and the third multicellular organism after *Caenorhabditis elegans* and *Drosophila melanogaster* whose genome was completely sequenced. To assist biological investigations and to define chromosomal structure, a coordinated effort to sequence the *A. thaliana* genome was initiated in late 1996, led by a consortium of researchers based mainly out of academic institutions in the USA, Europe and Japan (AGI). Separate teams within the consortium worked on different chromosomes, using distinct procedures. Today, the *Arabidopsis* community is a diverse group of scientists and involves universities, research institutes and private companies. Three papers presenting the DNA sequence of the gene-rich regions on chromosomes 1, 3 and 5 of A. thaliana were published first, followed by papers in which chromosomes 2 and 4 were described, altogether providing an overview on the *A. thaliana* sequence.

The initial identification of transcriptional units in the *A. thaliana* genome sequence was carried out largely by *ab initio* gene predictions, sequence homology, sequence motif analysis, and other non-experimental methods. **The Institute for Genomic Research (TIGR)** launched a reannotation effort, employing the latest annotation tools and resources and applying uniform annotation protocols across the entire genome, with the goal of improving annotation by refining gene structure and gene function assignments. The final TIGR genome reannotation release contains annotations for approximately 26,000 protein-coding genes. The completed sequence of a plant nuclear genome yielded a high number of insights, particularly when comparing it with the completed genomic sequences of other species available at the time, namely Caenorhabditis elegans and Drosophila melanogaster. A. thaliana had many families of new proteins but also lacked several common protein families, indicating that these sets of common proteins had undergone differential expansion and contraction in the three multicellular eukaryotes.

Now, let us turn our attention to sequences and the way they are dealt with using computational methods. We do this in **BioPython** using a random example from one of the reviewed protein sequence from *A. thaliana*. We have selected a gene, **FPF1** (*Flower Promoting Factor 1*), which modulates the competence to flowering of apical meristems. encodes a small protein of 12.6 kDa that regulates flowering and is involved in gibberellin signalling pathway. It is expressed in apical meristems immediately after the photoperiodic induction of flowering. Genetic interactions with flowering time and floral organ identity genes suggest that this gene may be involved in modulating the competence to flower. There are two other genes similar to FPF1, **FLP1** (At4g31380) and **FLP2** (no locus name yet, on BAC F8F16 on chr 4). This is so far, a plant-specific gene and is only found in long-day mustard, arabidopsis, and rice.

The gene can be accessed at NCBI (National Center for Biotechnology Information) by specifying the GeneID **832555**. It is located on Chromosome 5. Its FASTA sequence is given below:

> NC_003076.8:8541558-8542473 Arabidopsis thaliana chromosome 5 sequence CAACACCACAAGACCATAATTCA AGAATCAACACGTTGCTTCGCTAATTTATTCATCTTGTA GTT-ATTTCACCAATGGTCCCAAAAAAAATCATTGCTGG TTATAAT-TGTAAACAAATTCTATTTGTCACGTCAAAAAG ATATTTTTTTG-TCTTGATGTGGTTAGACCCTCTTCTATATAAT ATCATCTCCACACACAACTCTC-TTCTC AAAAAACACCAAAT CATTCAAGAAACCAAAACTAAGGA-GAGGAAGAAGAAGAT-CATGTCAGGCGTGTGGG TCTTCAAGAACGGAGTGATAAGACT-TGTGGAGAACCCTAACCAGTCAGGATCCGACACACAG AACC-GAAGGAAAGTGATGGTCTATTTACCGACAGGAG AAGTGG-TCTCATCTTACTCGACGCTCGAGCAGATCCTC CAGAGTCTTGGATGGGAGAGATACTT-TGGCGGCGGCG ACACAGATCTCCTCCAATTCCACAAACGCTCC-TCCATT GATCTCATTTCCCTGCCTAGAGATTTCACCAAATTCAA-CTCCGTTTACATGTACGATATCGTCGTCAA GAACCCTAATTAC-TTCCATGTCCGAGACTCCCATTGATTCTTCTTTTGTGTGTTGG TGTGTGTGTTGTGTTCGGTTGTTTTATAATTTA-GAGTTTGTTG GGTTTTGTTTTATAGTTGCATGTGCATGTTTTATTT-GTTTTGG-ACGTTGGAAGTTGTGTTACGTAATTCTGGTT-TGATGAAAGTTG AGATTAAGAAGGGACATGAATTAGGGAGAACTCAATTTCAAA ATTAATTGTTCT-CTTCCTTAGCATGATTGTTCTCTTCACTGGTT GTAATT-GAA AGATATTCTCTTTGGCTCAGTGGATCCAAATTGT TTTCT-ATCTTCTATTTAATTAATCAAAAGCCATATT AACGGA

The protein coded by this gene can be accessed at UniProt by specifying the ID **O23624**. The FASTA sequence is given below:

> sp|O23624|FPF1_ARATH Flowering-promoting factor 1 OS=Arabidopsis thaliana GN=FPF1 PE=2 SV=1 MSGVWVF KNGVIRLVENPNQSGSDTQNRRKVMVYLPTGEVVSSYS-TLEQILQSLGWERYFGGGDTDLLQFHKRSSIDLISLPRD-FTKFNSVYMYDIVVKNPNYFHVRDSH

It is assumed that the reader is familiar with the FASTA format and what each line means. If not, please refer to the books mentioned earlier.

A large part of bioinformatics work involves dealing with the many types of file formats designed to hold biological data. These files are loaded with interesting biological data, and a special challenge is parsing these files into a format so that you can manipulate them with some kind of programming language. However, the task of parsing these files can be frustrated by the fact that the formats can change quite regularly, and that formats may contain small subtleties which can break even the most well designed parsers.

We are now going to briefly introduce the **Bio.SeqIO** module – you can find out later. We show how to invoke and utilize the module. Interested reader should consult to understand all the attributes of the module.

```
from Bio import SeqIO
for seq_record in list(SeqIO.parse("flp1.fasta", "fasta"))[:5]:
   print(seq_record.id)
   print(repr(seq_record.seq))
   print(len(seq_record))
```

```
NC_003076.8:8541558-8542473
Seq('CAACACCACAAGACCATAATTCAAGAATCAACACGTTGC
TTCGCTAATTTTATT...GGA', SingleLetterAlphabet())
916
```

Notice that the FASTA format does not specify the alphabet, so **Bio.SeqIO** has defaulted to the rather *generic* SingleLetterAlphabet() rather than something DNA specific.

MULTIPLE SEQUENCE ALIGNMENT

By this we mean a collection of multiple sequences which have been aligned together – usually with the insertion of gap characters, and addition of leading or trailing gaps – such that all the sequence strings are the same length. Such an alignment can be regarded as a matrix of letters, where each row is held as a *SeqRecord* object internally in BioPython.

We will introduce the *MultipleSeqAlignment* object which holds this kind of data, and the Bio.AlignIO module for reading and writing them as various file formats (following the design of the Bio.SeqIO module). Note that both Bio.SeqIO and Bio.AlignIO can read and write sequence alignment files. The appropriate choice will depend largely on what you want to do with the data.

Parsing or Reading Sequence Alignments

We have two functions for reading in sequence alignments, Bio.AlignIO.read() and Bio.AlignIO.parse() which following the convention introduced in Bio.SeqIO for files containing one or multiple alignments respectively. Using Bio.AlignIO.parse() will return an *iterator* which gives *MultipleSeqAlignment* objects. Iterators are typically used in a *for* loop. In many situations, you will be dealing with files which contain only a single alignment. In this case, you should use the **Bio.AlignIO.read()** function which returns a single *MultipleSeqAlignment* object.

Both functions expect two mandatory arguments:

- The first argument is a *handle* to read the data from, typically an open file, or a filename.
- The second argument is a lower case string specifying the *alignment format*.

As in **Bio.SeqIO** we don't try and guess the file format for you! Please visit the following website for a full listing of supported formats.

http://biopython.org/wiki/AlignIO

We take an example of seed alignment for the **NolB** (**PF17398**) PFAM entry. This can be downloaded from Pfam website given below, by specifying the Pfam ID above. Family members of NolB are involved in *Nodulation* (nodule development in plants).

http://pfam.xfam.org/

from Bio import AlignIO

alignment = AlignIO.read("PF17398_seed.txt", "stockholm")

print(alignment)

SingleLetterAlphabet() alignment with 7 rows and 153 columns

SPS-FSEQAQFERALAHAADSLKNDAASAPSRV-PV--PPAMAL... KEG Q989P6_RHILO/21-167

SSS-FGEQAQFERALAQAADSMKNDTASTPVRTAPI--SPPMDV...

KEG NOLB_SINFN/17-164

ERSACGSQAQFECALVEAVASIKNDGPSATAGAGPV--APVLQV... KQG A0A090F432_9RHIZ/61-210

SPAAAGEQARFEQSLAQVAP---NQGSAFPTANPPAATPPILEV... SAG Q89TP9_BRADU/25-170

SAAALGDQAQFQRSLVQAASSVKNDASSATAGVAPV--PPTLEV... SAG G6YH87_9RHIZ/21-171

SPS-FGEQAQFEHALAQAAASMKNDTASGPATSAPI--PPQLDV... KEG H0HXE4_9RHIZ/21-170

SPS-LGELAQFERALVQAAASTKNDTISRPESVASV--SPHLDV... KQG I4Z467_9RHIZ/21-164

 # accessing individual records in the MSA (multiple sequence alignment)

for record **in** alignment:

  print(record)

ID: Q989P6_RHILO/21-167

Name: Q989P6_RHILO

Description: Q989P6_RHILO/21-167

Number of features: 0

/start=21

/end=167

/accession=Q989P6.1

Seq('SPS-FSEQAQFERALAHAADSLKNDAASAPSRV-PV--PPAMALQRAATQMSPL...KEG', SingleLetterAlphabet())

ID: NOLB_SINFN/17-164

Name: NOLB_SINFN

Description: NOLB_SINFN/17-164

Number of features: 0

/start=17

/end=164

/accession=P55713.1

Seq('SSS-FGEQAQFERALAQAADSMKNDTASTPVRTAPI--

SPPMDVHRAAPTS-PL...KEG', SingleLetterAlphabet())

ID: A0A090F432_9RHIZ/61-210

Name: A0A090F432_9RHIZ

Description: A0A090F432_9RHIZ/61-210

Number of features: 0

/start=61

/end=210

/accession=A0A090F432.1

Seq('ERSACGSQAQFECALVEAVASIKNDGPSATAGAGPV--APVLQVPRATSETSPL...KQG', SingleLetterAlphabet())

ID: Q89TP9_BRADU/25-170

Name: Q89TP9_BRADU

Description: Q89TP9_BRADU/25-170

Number of features: 0

/start=25

/end=170

/accession=Q89TP9.1

Seq('SPAAAGEQARFEQSLAQVAP---NQGSAFPTANPPAATPPILEVQRTNVLASPP...SAG', SingleLetterAlphabet())

ID: G6YH87_9RHIZ/21-171

Name: G6YH87_9RHIZ

Description: G6YH87_9RHIZ/21-171

Number of features: 0

/start=21

/end=171

/accession=G6YH87.1

Seq('SAAALGDQAQFQRSLVQAASSVKNDASSATAGVAPV--PPTLEVQRAVTQANPL...SAG', SingleLetterAlphabet())

ID: H0HXE4_9RHIZ/21-170

Name: H0HXE4_9RHIZ

```
Description: H0HXE4_9RHIZ/21-170
Number of features: 0
/start=21
/end=170
/accession=H0HXE4.1
Seq('SPS-FGEQAQFEHALAQAAASMKNDTASGPATSAPI--PPQLDVQRATAQSSPL...KEG', SingleLetterAlphabet())
ID: I4Z467_9RHIZ/21-164
Name: I4Z467_9RHIZ
Description: I4Z467_9RHIZ/21-164
Number of features: 0
/start=21
/end=164
/accession=I4Z467.1
Seq('SPS-LGELAQFERALVQAAASTKNDTISRPESVASV--SPHLDVQRTTAQTNPW...KQG', SingleLetterAlphabet())
# getting the length or the number of residues in each sequence
sizes = [len(rec) for rec in alignment]
sizes
[153, 153, 153, 153, 153, 153, 153]
```

FREQUENCY ANALYSIS

Perhaps the most basic analysis to perform on a sequence is to determine the frequency of occurrence of its component elements. In the case of biological sequences, the ratio of the four nucleotides or twenty amino acids in a given region of nucleic acid or protein sequence can give us important structural and functional clues. Similarly, for the relative proportions of dimers, trimers, and higher order oligomers; i.e. pairs, triplets, etc. of nucleotides or amino acids.

For example, given the fact that A-T base pairing involves two hydrogen bonds, while G-C pairing involves three, DNA with a low G+C content will separate into its two component strands at lower temperatures than that with a higher G+C ratio. In fact, it has been demonstrated that *thermophilic* microorganisms living in hot springs and deepsea vents tend to have genomes with higher G+C values than those living in cooler environments. Similarly, because of various steps that require the binding and subsequent dissociation of complementary DNA strands, genomes with a high G + C content are more difficult to sequence. Higher G + C content has also been associated with more gene-dense regions of mammalian genomes.

Oligomer frequency can be used to distinguish coding and noncoding regions. The coding regions (exons) in the DNA of higher organisms such as humans are typically short (averaging 150 nucleotides in length) and widely separated. Current estimates are that only 1-3% of human DNA is comprised of coding regions. In order to determine protein sequences based upon the DNA sequences of human genes, these coding regions need to be located.

Let us quickly look at a simple example of nucleotide sequences. We choose an example from **BioPython** itself. The file has sequences from the plant orchid. Let us evaluate the GC% of nucleotide sequences as an example. We are going to use an example from BioPython documentation for illustration. The sequences are from **Lady Slipper Orchids**. Orchids are in the *Orchidaceae* family and the *Cypripedioideae* sub-family and are made up of 5 genera: *Cypripedium, Paphiopedilum, Phragmipedium, Selenipedium and Mexipedium.* You can extract the sequences from NCBI website. The file here contains 94 hits; you may get more hits in your search. The file has been named *ls_orchid.fasta*.

```
from Bio.SeqUtils import GC
gc_values = sorted(GC(rec.seq) for rec in SeqIO.
parse('ls_orchid.fasta', 'fasta'))
print(len(gc_values))
94
```

Human gene mutations have also been observed more frequently at certain dinucleotides than others. It is postulated that this *dimer bias* is related in part to the ability of certain amino acids to substitute for one another in a protein without significant loss of function and to their corresponding edit distance in the triplet genetic code.

The *frequency* of various amino acids in a protein sequence has consequences for the protein's behavior when dissolved in water (as in the cell's cytoplasm) or in the lipid environment of the cell membrane. Methods of distinguishing hydrophobic and hydrophilic regions of proteins have been used to predict the gross three-dimensional structure of proteins, their likely home within the cell, and their broad functional characteristics.

The frequencies of short oligomers can also be used to detect a variety of important biological features in a sequence. A few cases in point are given below.

- Of a short-range effect, the region just before the splice site at the end of an intron typically has high C + T content. As with many such sequence patterns, this is detectable with statistical methods, but the pattern is not clear enough to be detected without also detecting many false positives.
- An even shorter pattern is the near-100% occurrence of the dinucleotide GT as the first two bases of an intron, and AG as the last two bases.
- Another example of local sequence correlation is the *ribosome binding site*, which occurs just before the start codon that initiates protein translation. The sequence of this site, which varies in different organisms, shows clear preferences for some bases over others in particular positions. Recent evidence indicates that in addition to these positional preferences there are strong correlations between adjacent positions in the site. This makes sense from a biological standpoint, since the ribosome is a large molecular complex that binds to a region of DNA, not to individual bases. Sequence modeling techniques such as Markov chains are especially effective at capturing these local dependencies.

More subtle patterns can be detected by looking for long-range correlations between sites that, although distant in the sequence, are in close proximity in the three-dimensional environment of the cell. For example, transcriptional regulation occurs via proteins that bind to patterns in the DNA sequence at a considerable distance from the transcribed region. Regulatory sequences known as "*promoters*" and "*enhancers*" can occur anywhere from a few bases upstream of the start of transcription to thousands of bases away. They can even occur within the transcribed region itself. Sequence analysis software is just beginning to reliably identify these signals.

MEASURING HOMOLOGY

Two proteins are **homologous** if they have similar three-dimensional structures; similar shape usually implies similar function. Because protein sequence and three-dimensional structure are closely related, we can take this a step further and say that if two protein sequences are sufficiently similar, then they are likely to have similar functions. This observation leads to a very useful method for determining the function of an unknown gene: by matching the predicted amino acid sequence of an uncharacterized but sequenced gene to the sequence of a known gene, we can infer the fimction of the new gene from the known one. The underlying assumption is that two sufficiently similar genes probably evolved from a common ancestor, and the degree of similarity visible today is a function of how long ago the sequences (and the organisms from which they were taken) diverged. Nucleic acid sequences are also compared, although the redundancy in the genetic code (whereby multiple codons map to the same amino acid) means that the corresponding amino acid sequences might be much more similar than the DNA sequences for any given pair of genes.

The process of comparing two sequences involves *aligning* them and inserting *gaps* in one or both sequences in order to produce an *optimal score*. An alignment is scored by adding up the match scores and subtracting penalties for the gaps. Match scores are contained in a *matrix* that assigns a score for each of the 400 possible amino acid pairs. For DNA sequence alignment, match scores usually just assign a fixed score for an identical nucleotide match and a different score for a mismatch.

Given a *scoring matrix* and a *gap penalty*, the optimal alignment for two sequences can be computed by dynamic programming in $O(n^2)$ time. The most general formulation of the algorithm allows one to align just a subsequence of one input sequence to a subsequence of the other; the entire sequences do not have to be similar.

When comparing and aligning protein sequences, it is important to consider that certain amino acids can substitute for others in a protein without dramatically altering the molecule's three-dimensional structure or function. Thus, *mismatch penalties* in protein alignment algorithms usually employ some sort of *scoring matrix*, a table which indicates the penalties associated with each possible amino acid substitution. Since there is no good way to measure the substitutability of one amino acid for another *a priori*, these scoring matrices have been empirically derived in the past based on a bootstrapping process which postulates an alignment of sequences with a putative evolutionary "distance", and then generates a score based on the frequency of substitution of alternative residues at each position.

The most popular scoring matrices for general protein searching are the BLOSUM set of Henikoff and Henikoff, and the PAM matrices of Dayhoff. Since, there are 20 amino acids the substitution matrix is 20 x 20. This matrix indicates the *log-odds* of a substitution. The matrices may be presented in different arrangements and so it is necessary to first define the alphabet that is associated with the use of a matrix. In this case the alphabet is 'ARNDCQEGHILKMFPSTWYV'. The BLOSUM50 matrix is shown in Table 1. Each row and column is associated with a letter. So the *log-odds* of an 'A' changing to an 'R' is -2.

The odds of an event occurring is based on the probability of the event normalized by the probability of the constituents existing. The range of the values for the odds can be large and that is one of the reason that the log-odds are used. Negative values indicate that the odds were less than 1 and positive values indicated that the odds were greater than 1. Thus, *positive values* in the table are events that are *likely to occur* with larger values indicating a larger chance. As seen all of the events in which an item remains unchanged (values down the diagonal of the matrix) are positive and are the largest values.

It is also important to keep in mind that sequence alignment does not always imply functional similarity. This becomes an especially important caveat when the alignment score is relatively weak, indicating only distant similarity. The power and speed of sequence alignment programs makes it easier than ever to find weak homologies that might not represent true evolutionary relationships.

Table 5.1: BLOSUM50 matrix

	A	R	N	D	C	Q	E	G	H	I	L	K	M	F	P	S	T	W	Y	V
A	5	-2	-1	-2	-1	-1	-1	0	-2	-1	-2	-1	-1	-3	-1	1	0	-3	-2	-2
R	-2	7	-1	-2	-1	1	0	-3	0	-4	-3	3	-2	-3	-3	-1	-1	-3	-1	-1
N	-1	-1	7	2	-2	0	0	0	1	-3	-4	0	-2	-4	-2	-1	0	-4	-2	-2
D	-2	-2	2	8	-4	0	2	-1	-1	-4	-4	-1	-4	-5	-1	0	-1	-5	-3	-3
C	-1	-4	-2	-4	13	-3	-3	-3	-3	-2	-2	-3	-2	-2	-4	-1	-1	-5	-3	-3
Q	-1	-1	0	0	-3	7	2	-2	1	-3	-2	2	0	-4	-1	0	-1	-1	-1	-1
E	-1	0	0	2	-3	2	6	-3	0	-4	-3	1	-2	-3	-1	-1	-1	-3	-2	-2
G	0	-3	0	-1	-3	-2	-3	8	-2	-4	-4	-2	-3	-4	-2	0	-2	-3	-3	-3
H	-2	0	1	-1	-3	1	0	-2	10	-4	-3	0	-1	-1	-2	-1	-2	-3	-1	-1
I	-1	-4	-3	-4	-2	-3	-4	-4	-4	5	2	-3	2	0	-3	-3	-1	-3	-1	-1
L	-2	-3	-4	-4	-2	-2	-3	-4	-3	2	5	-3	3	1	-4	-3	-1	-2	-1	-1
K	-1	3	0	-1	-3	2	1	-2	0	-3	-3	6	-2	-4	-1	0	-1	-3	-2	-2
M	-1	-2	-2	-4	-2	0	-2	-3	-1	2	3	-2	7	0	-3	-2	-1	-1	0	0
F	-3	-3	-4	-5	-2	-4	-3	-4	-1	0	1	-4	0	8	-4	-3	-2	1	4	4
P	-1	-3	-2	-1	-4	-1	-1	-2	-2	-3	-4	-1	-3	-4	10	-1	-1	-4	-3	-3
S	1	-1	1	0	-1	0	-1	0	-1	-3	-3	0	-2	-3	-1	5	2	-4	-2	-2
T	0	-1	0	-1	-1	-1	-1	-2	-2	-1	-1	-1	-1	-2	-1	2	5	-3	-2	-2
W	-3	-3	-4	-5	-5	-1	-3	-3	-3	-3	-2	-3	-1	-1	-4	-4	-3	-15	2	2
Y	-2	-1	-2	-3	-3	-1	-2	-3	2	-1	-1	-2	0	4	-3	-2	-2	2	8	8
V	0	-3	-3	-4	-1	3	-3	-4	-4	4	1	-3	1	-1	-3	-2	0	-3	-1	-1

We present an example of MSA and then go on to describe the different algorithms. Let us now load and parse some precomputed MSA available from Pfam. We chose, for illustration, MSA of protein family **PF17398**, downloaded from

http://pfam.sanger.ac.uk

```
from Bio import AlignIO
alignment = AlignIO.read("PF17398_seed.txt", "stockholm")
```

Let us print summary of the alignment

```
print(alignment)
```

```
SingleLetterAlphabet() alignment with 7 rows and 153 columns
SPS-FSEQAQFERALAHAADSLKNDAASAPSRV-PV--PPAMAL...
KEG Q989P6_RHILO/21-167
SSS-FGEQAQFERALAQAADSMKNDTASTPVRTAPI--SPPMDV...
KEG NOLB_SINFN/17-164
ERSACGSQAQFECALVEAVASIKNDGPSATAGAGPV--APVLQV...
KQG A0A090F432_9RHIZ/61-210
SPAAAGEQARFEQSLAQVAP---NQGSAFPTANPPAATPPILEV...
SAG Q89TP9_BRADU/25-170
SAAALGDQAQFQRSLVQAASSVKNDASSATAGVAPV--PPTLEV...
SAG G6YH87_9RHIZ/21-171
```

SPS-FGEQAQFEHALAQAAASMKNDTASGPATSAPI--PPQLDV... KEG H0HXE4_9RHIZ/21-170

SPS-LGELAQFERALVQAAASTKNDTISRPESVASV--SPHLDV... KQG I4Z467_9RHIZ/21-164

You'll notice in the above output the sequences have been truncated. We could instead write our own code to format this as we please by iterating over the rows as *SeqRecord* objects:

print("Alignment length %i" % alignment.get_alignment_length())

Alignment length 153

for record in alignment:

print("%s - %s" % (record.seq, record.id))

SPS-FSEQAQFERALAHAADSLKNDAASAPSRV-PV--PPAMALQRAATQMSPLGDRVLQTISAMYPDRAVNSAALDQQVGLFKGALPRP-QKLPAEGD-AGTGTSGVPQGRHDFETMIAGLRDMYNGVTQVALVSKGVSGITSSVNKLLKEG - Q989P6_RHILO/21-167

SSS-FGEQAQFERALAQAADSMKNDTASTPVRTAPI--SPPMDVHRAAPTS-PLEDRVLQTISSICPDSIVAAAAPNHKAALISGAPPGPSQKLPVEGG-AGTERLGIPQGGHDFDVMVAGLRDLYNGVTQVALVSKGISGITSSVNKLLKEG - NOLB_SINFN/17-164

ERSACGSQAQFECALVEAVASIKNDGPSATAGAGPV--APVLQVPRATSETSPLGDRVLQTLSSLYRDNAVTPATPDQEVPVVKDVLLGPAEQSPSQAG-SGMRMSGAPPGRENYEAMIASLREVYSAVTEVSLVTKSISGATSSVNTLIKQG - A0A090F432_9RHIZ/61-210

SPAAAGEQARFEQSLAQVAP---NQGSAFPTANPPAATPPILEVQRTNVLASPPGDRILQTLSGIYQGRAVSPAIAP---AVVNGVQPGPAAQPLLPADNAGANA-AVKPVGADFESMMTNLRDVYKDVVQVSLVSKGTSAVSSSLNKLLSAG - Q89TP9_BRADU/25-170

SAAALGDQAQFQRSLVQAASSVKNDASSATAGVAPV--PPTLEVQRAVTQANPLGDRVLQTLSSMHQHNAVAPPALGREVPLVTGAPPGPAAQLPLQARAAGVQTPSAPQGGKDFEAMVAGLRDVYNDVTQVSLVCKGIGSASSSVNKLVSAG - G6YH87_9RHIZ/21-171

SPS-FGEQAQFEHALAQAAASMKNDTASGPATSAPI--PPQLD-VQRATAQSSPLGDRVLQTISSMYRDNAVTSAALDHQVGLAKDG-

LPGPAQKLPVEGGAAGAQMSGVPQAGHDFESMIADLRDLYN-GATQVALISKGVSGITSSVNKLLKEG - H0HXE4_9RHIZ/21-170

SPS-LGELAQFERALVQAAASTKNDTISRPESVASV--SPHLDVQR TTAQTNPWGDRVFEMISSISRFDSVTAAVHDNQLGLEKSALSGPA QKLPAEGAAPGT------PIGQDFGAVIAELRDLYYGVTEVALISKGV SGITSSVNTLIKQG - I4Z467_9RHIZ/21-164

You could also use the alignment object's *format* method to show it in a particular file format. Let us do some *manipulations* of alignments.

Slicing Alignments

First of all, in some senses the alignment objects act like a *Python* list of *SeqRecord* objects (the rows). With this model in mind hopefully the actions of *len()* (the number of rows) and iteration (each row as a *SeqRecord*) make sense:

```
print("Number of rows: %i" % len(alignment))
Number of rows: 7
for record in alignment:
    print("%s - %s" % (record.seq, record.id))
```

SPS-FSEQAQFERALAHAADSLKNDAASAPSRV-PV--PPAMALQ RAATQMSPLGDRVLQTISAMYPDRAVNSAALDQQVGLFKGALP RP-QKLPAEGD-AGTGTSGVPQGRHDFETMIAGLRDMYNGVTQV ALVSKGVSGITSSVNKLLKEG - Q989P6_RHILO/21-167

SSS-FGEQAQFERALAQAADSMKNDTASTPVRTAPI--SPPMDVHRAAPTS-PLEDRVLQTISSICPDSIVAAAAPNHKAALIS GAPPGPSQKLPVEGG-AGTERLGIPQGGHDFDVMVAGLRDLYNG VTQVALVSKGISGITSSVNKLLKEG - NOLB_SINFN/17-164

ERSACGSQAQFECALVEAVASIKNDGPSATAGAGPV--APVLQV PRATSETSPLGDRVLQTLSSLYRDNAVTPATPDQEVPVVKDVL LGPAEQSPSQAG-SGMRMSGAPPGRENYEAMIASLREVYSAV TEVSLVTKSISGATSSVNTLIKQG - A0A090F432_9RHIZ/61-210 SPAAAGEQARFEQSLAQVAP---NQGSAFPTANPPAATP PILEVQRTNVLASPPGDRILQTLSGIYQGRAVSPAIAP---AVVNGVQPGPAAQPLLPADNAGANA-AVKPVGADFESMMT NLRDVYKDVVQVSLVSKGTSAVSSSLNKLLSAG - Q89TP9_

BRADU/25-170

SAAALGDQAQFQRSLVQAASSVKNDASSATAGVAPV--PPTLEVQ RAVTQANPLGDRVLQTLSSMHQHNAVAPPALGREVPLVTGAPPG PAAQLPLQARAAGVQTPSAPQGGKDFEAMVAGLRDVYNDVTQ VSLVCKGIGSASSSVNKLVSAG - G6YH87_9RHIZ/21-171

SPS-FGEQAQFEHALAQAAASMKNDTASGPATSAPI--PPQLDVQR ATAQSSPLGDRVLQTISSMYRDNAVTSAALDHQVGLAKDGLPGP AQKLPVEGGAAGAQMSGVPQAGHDFESMIADLRDLYNGATQV ALISKGVSGITSSVNKLLKEG - H0HXE4_9RHIZ/21-170

SPS-LGELAQFERALVQAAASTKNDTISRPESVASV--SPHLDVQR TTAQTNPWGDRVFEMISSISRFDSVTAAVHDNQLGLEKSALSGPA QKLPAEGAAPGT------PIGQDFGAVIAELRDLYYGVTEVALISKGV SGITSSVNTLIKQG - I4Z467_9RHIZ/21-164

You can also use the list-like *append* and *extend* methods to add more rows to the alignment (as *SeqRecord* objects). Keeping the *list* metaphor in mind, simple slicing of the alignment should also make sense. Select some of the rows giving back another alignment object:

print(alignment)

SingleLetterAlphabet() alignment with 7 rows and 153 columns

SPS-FSEQAQFERALAHAADSLKNDAASAPSRV-PV--PPAMAL... KEG Q989P6_RHILO/21-167

SSS-FGEQAQFERALAQAADSMKNDTASTPVRTAPI--SPPMDV... KEG NOLB_SINFN/17-164

ERSACGSQAQFECALVEAVASIKNDGPSATAGAGPV--APVLQV... KQG A0A090F432_9RHIZ/61-210

SPAAAGEQARFEQSLAQVAP---NQGSAFPTANPPAATPPILEV... SAG Q89TP9_BRADU/25-170

SAAALGDQAQFQRSLVQAASSVKNDASSATAGVAPV--PPTLEV... SAG G6YH87_9RHIZ/21-171

SPS-FGEQAQFEHALAQAAASMKNDTASGPATSAPI--PPQLDV... KEG H0HXE4_9RHIZ/21-170

SPS-LGELAQFERALVQAAASTKNDTISRPESVASV--SPHLDV... KQG I4Z467_9RHIZ/21-164

print(alignment[3:7])

```
SingleLetterAlphabet() alignment with 4 rows and 153 columns
SPAAAGEQARFEQSLAQVAP---NQGSAFPTANPPAATPPILEV...
SAG Q89TP9_BRADU/25-170
SAAALGDQAQFQRSLVQAASSVKNDASSATAGVAPV--PPTLEV...
SAG G6YH87_9RHIZ/21-171
SPS-FGEQAQFEHALAQAAASMKNDTASGPATSAPI--PPQLDV...
KEG H0HXE4_9RHIZ/21-170
SPS-LGELAQFERALVQAAASTKNDTISRPESVASV--SPHLDV...
KQG I4Z467_9RHIZ/21-164
```

What if you wanted to select by column? Those of you who have used the **NumPy** matrix or *array* objects won't be surprised at this - you use a **double index**.

```
print(alignment[2, 6])
S
```

Using two integer indices pulls out a single letter, short hand for this:

```
print(alignment[2].seq[6])
S
```

You can pull out a single column as a string like this:

```
print(alignment[:, 6])
EESEDEE
```

You can also select a range of columns. For example, to pick out those same three rows we extracted earlier, but take just their first six columns:

```
print(alignment[3:6, :6])
SingleLetterAlphabet() alignment with 3 rows and 6 columns
SPAAAG Q89TP9_BRADU/25-170
SAAALG G6YH87_9RHIZ/21-171
SPS-FG H0HXE4_9RHIZ/21-170
```

Leaving the first index as : means take all the rows:

```
print(alignment[:, :6])
SingleLetterAlphabet() alignment with 7 rows and 6 columns
SPS-FS Q989P6_RHILO/21-167
```

SSS-FG NOLB_SINFN/17-164

ERSACG A0A090F432_9RHIZ/61-210

SPAAAG Q89TP9_BRADU/25-170

SAAALG G6YH87_9RHIZ/21-171

SPS-FG H0HXE4_9RHIZ/21-170

SPS-LG I4Z467_9RHIZ/21-164

Again, you can slice to get everything after the 120th column:

```
print(alignment[:, 9:])
```

SingleLetterAlphabet() alignment with 7 rows and 144 columns

QFERALAHAADSLKNDAASAPSRV-PV--PPAMALQRAATQMSP... KEG Q989P6_RHILO/21-167

QFERALAQAADSMKNDTASTPVRTAPI--SPPMDVHRAAPTS-P... KEG NOLB_SINFN/17-164

QFECALVEAVASIKNDGPSATAGAGPV--APVLQVPRATSETSP... KQG A0A090F432_9RHIZ/61-210

RFEQSLAQVAP---NQGSAFPTANPPAATPPILEVQRTNVLASP... SAG Q89TP9_BRADU/25-170

QFQRSLVQAASSVKNDASSATAGVAPV--PPTLEVQRAVTQANP... SAG G6YH87_9RHIZ/21-171

QFEHALAQAAASMKNDTASGPATSAPI--PPQLDVQRATAQSSP... KEG H0HXE4_9RHIZ/21-170

QFERALVQAAASTKNDTISRPESVASV--SPHLDVQRTTAQTNP... KQG I4Z467_9RHIZ/21-164

Now, the interesting thing is that addition of alignment objects works by column. This lets you do this as a way to remove a block of columns:

```
edited = alignment[:, :6] + alignment[:, 9:]
print(edited)
```

SingleLetterAlphabet() alignment with 7 rows and 150 columns

SPS-FSQFERALAHAADSLKNDAASAPSRV-PV--PPAMALQRA... KEG Q989P6_RHILO/21-167

SSS-FGQFERALAQAADSMKNDTASTPVRTAPI--SPPMDVHRA... KEG NOLB_SINFN/17-164

ERSACGQFECALVEAVASIKNDGPSATAGAGPV--APVLQVPRA... KQG A0A090F432_9RHIZ/61-210

SPAAAGRFEQSLAQVAP---NQGSAFPTANPPAATPPILEVQRT... SAG Q89TP9_BRADU/25-170

SAAALGQFQRSLVQAASSVKNDASSATAGVAPV--PPTLEVQRA... SAG G6YH87_9RHIZ/21-171

SPS-FGQFEHALAQAAASMKNDTASGPATSAPI--PPQLDVQRA... KEG H0HXE4_9RHIZ/21-170

SPS-LGQFERALVQAAASTKNDTISRPESVASV--SPHLDVQRT... KQG I4Z467_9RHIZ/21-164

Another common use of alignment addition would be to combine alignments for several different genes into a meta-alignment. Watch out though - the identifiers need to match up. You may find it helpful to first sort the alignment rows alphabetically by id:

edited.sort()

print(edited)

SingleLetterAlphabet() alignment with 7 rows and 150 columns

ERSACGQFECALVEAVASIKNDGPSATAGAGPV--APVLQVPRA... KQG A0A090F432_9RHIZ/61-210

SAAALGQFQRSLVQAASSVKNDASSATAGVAPV--PPTLEVQRA... SAG G6YH87_9RHIZ/21-171

SPS-FGQFEHALAQAAASMKNDTASGPATSAPI--PPQLDVQRA... KEG H0HXE4_9RHIZ/21-170

SPS-LGQFERALVQAAASTKNDTISRPESVASV--SPHLDVQRT... KQG I4Z467_9RHIZ/21-164

SSS-FGQFERALAQAADSMKNDTASTPVRTAPI--SPPMDVHRA... KEG NOLB_SINFN/17-164

SPAAAGRFEQSLAQVAP---NQGSAFPTANPPAATPPILEVQRT... SAG Q89TP9_BRADU/25-170

SPS-FSQFERALAHAADSLKNDAASAPSRV-PV--PPAMALQRA... KEG Q989P6_RHILO/21-167

Note that you can only add two alignments together if they have the same number of rows.

Alignment as arrays

Depending on what you are doing, it can be more useful to turn the alignment object into an array of letters – and you can do this with **NumPy**:

```
import numpy as np
align_array = np.array([list(rec) for rec in alignment], np.character)
print("Array shape %i by %i" % align_array.shape)
Array shape 7 by 153
```

If you will be working heavily with the columns, you can tell **NumPy** to store the array by column (as in Fortran) rather then its default of by row (as in C):

```
align_array = np.array([list(rec) for rec in alignment], np.character, order="F")
```

GLOBAL AND LOCAL ALIGNMENTS

There are two types of alignments, which can be done given a set of sequences.

- The first is that the beginning and the end of the genes are known and so the two strings that are to be aligned have a definite beginning and ending. This is a global alignment since the process aligns the entirety of both strings. This is named *Needleman-Wunsch* alignment after the creators.
- The second case is that the user has two strings of DNA and inside of these are regions of interest but the beginning and ending of these regions is not known. Thus, the user is interested in finding a portion of one string that aligns with a portion of the other. This is called local alignment since the alignment result generally uses only a part of each string. This is also called *Smith-Waterman* alignment.

Let us look in some detail about the alignment details at an algorithmic level.

Global Comparison - Basic Algorithm

By alignment we mean insertion of a number of spaces between two sequences, in arbitrary locations, so that two sequences end up with same size. With same size, we can place the two sequences above one another creating a correspondence between characters. Given an alignment, we assign a score as discussed in the previous section.

It basically consists of solving an instance of a problem by taking advantage of already computed solutions for smaller instances of the same problem. Given two sequences s and t, instead of determining the similarity between s and t as whole sequences only, we build up the solution by determining all similarities between arbitrary *prefixes* of the two sequences. We start with the shorter prefixes and use previously computed results to solve the problem for larger prefixes.

Let m be the size of s and n the size of t. There are $m + 1$ possible prefixes of s and $n + 1$ prefixes of t, including the empty string. Thus, we can arrange our calculations in an $(m + 1) \times (n + 1)$ array where entry (i, j) contains the similarity between $s[1..i]$ and $t[1..j]$.

Figure 1 shows the array corresponding to s=AAAC and t=AGC. We placed s along the left margin and t along the top to indicate the prefixes more easily. Notice that the first row and the first column are initialized with multiples of the space penalty (-2 in our case). This is because there is only one alignment possible if one of the sequences is empty: Just add as many spaces as there are characters in the other sequence. The score of this alignment is $-2k$, where k is the length of the nonempty sequence. Hence, filling the first row and column is easy.

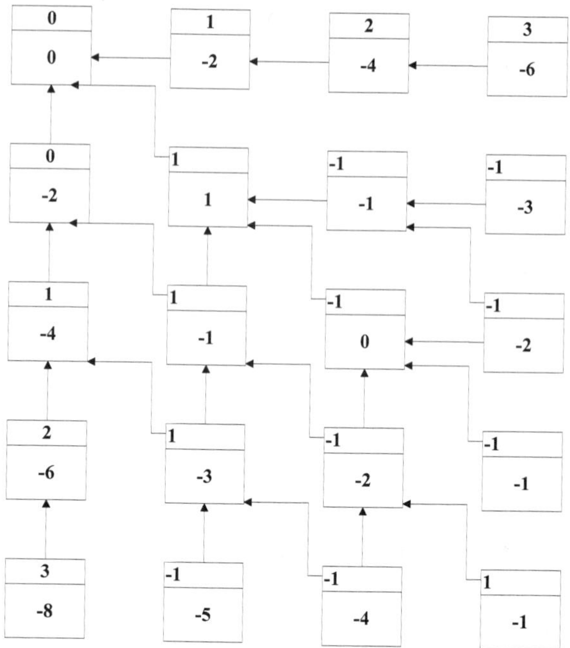

Figure 1. Bi-dimensional array for computing optimal alignments. The value in the upper right corner of cell (i, j) indicates whether $s[i] = t[j]$. Indices of rows and columns start at zero.

Now let us concentrate on the other entries. The key observation here is that we can compute the value for entry (i, j) looking at just three previous entries: those for $(i - 1, j)$, $(i - 1, j - 1)$ and $(i, j - 1)$. The reason is that there are just three ways of obtaining an alignment between $s[1..i]$ and $t[1..j]$, and each one uses one of these previous values. In fact, to get an alignment for $s[1..i]$ and $t[1..j]$, we have the following three choices:

- Align $s[1..i]$ with $t[1..j - 1]$ and match a space with $t[j]$, or
- Align $s[1..i - 1]$ with $t[1..j - 1]$ and match $s[i]$ with $t[j]$, or
- Align $s[1..i - 1]$ with $t[1..j]$ and match $s[i]$ with a space.

These possibilities are exhaustive because we cannot have two spaces paired in the last column of the alignment. Scores of the best alignments between smaller prefixes are already stored in the array if we choose an appropriate order in which to compute the entries. As a consequence, the similarity sought can be determined by the formula:

$sim(\text{s}[1..\text{i}], \text{t}[1..\text{j}]) = \max$

where $p(i, j)$ is +1 if $s[i] = t[j]$ and -1 if $s[i]$ $t[j]$. The values of $p(i, j)$ are written in the upper left corners of the boxes in Figure 1. Let us denote the array by a. Then the above equation can be rewritten as:

$a[i, j]$ = max

Filling in the array either row by row, left to right on each row, or column by column, top to bottom on each column, suffices. Any other order that makes sure $a[i, j - 1]$, $a[i - 1, j - 1]$, and $a[i - 1, j]$ are available when a[i, j] must be computed is fine, too. Finally, we drew arrows in Figure 3.1 to indicate where the maximum value comes from according to above equation. Therefore, there is only one way of getting this maximum value, namely, coming from entry (1, 1), and that is what the arrows show.

Optimal alignments

We have discussed a *generic* alignment algorithm. Now, let us see how we can arrive at an *optimal* alignment. In other words, we need to compute **similarity** between two sequences. You can focus on arrows in Figure 1 for understanding this concept.

All we need to do is start at entry (m, n) and follow the arrows until we get to (0, 0). Each arrow used will give us one column of the alignment. In fact, consider an arrow leaving entry (i, j). If this arrow is horizontal, it corresponds to a column with a space in s matched with $t[j]$; if it is vertical, it corresponds to $s[i]$ matched with a space in t; finally, a diagonal arrow means $s[i]$ matched with $t[j]$. Observe that the first sequence, s, is always placed along the vertical edge. Thus, an optimal alignment can be easily constructed from right to left if we have the matrix a computed by the basic algorithm. It is not necessary to implement the arrows explicitly - a simple test can be used to choose the next entry to visit.

Figure 2 shows a recursive algorithm for determining an optimal alignment given the matrix a and sequences s and t. The call *Align(m, n, len)* will construct an optimal alignment. The answer will be given in a pair of vectors *align-s* and *align-t* that will hold in the positions 1. .*len* the aligned characters, which can be either spaces or symbols from the sequences. The variables *align-s* and *align-t* are treated as globals in the code. The length of the alignment is also returned by the algorithm in the parameter *len*. Note that $\max(|s|, |t|) \leq len \leq m + n$.

Algorithm *Align*

input: indices *i, j* , array *a* given by algorithm *Similarity*

output: alignment in *aligns, align-t,* and length in *len*

if $i = 0$ **and** $j = 0$ **then**

$len \leftarrow 0$

else if $i > 0$ **and** $a[i, j] = a[i - 1, j] + g$ **then**

Align(i - 1, j , len)

$len \leftarrow len + 1$

align-s[len] ← *s[i]*

align-t[len] ← -

else if $i > 0$ **and** $j > 0$ **and** $a[i, j] = a[i - 1, j -1] + p(i, j)$ **then**

Align(i - 1, j - 1, len)

$len \leftarrow len + 1$

align-s[len] ← *s[i]*

align-t[len] ← *t[j]*

else *i has to be* $j > 0$ *anda*$[i, j] = a[i, j - 1] + g$

Align(i, j - 1, len)

$len \leftarrow len + 1$

aligns[len] ← -

align-t[len] ← *t[j]*

Figure 2. Algorithm for aligning two sequences

As a result, the optimal alignment returned by this algorithm has the following general characteristic: When there is choice, a column with a space in t has precedence over a column with two symbols, which in turn has precedence over a column with a space in s. For instance, when aligning s = AT AT with t = TATA, we get

-ATAT

TATA-

rather than

ATAT-

-TATA

which is the other optimal alignment for these two sequences. Similarly, when aligning s = AA with t = AAAA, we get

--AA

AAAA

although there are five other optimal alignments. This alignment is sometimes referred to as the *upmost* alignment because it uses the arrows higher up in the matrix. To reverse these preferences, we would reverse the order of the if statements in the code, obtaining the *downmost* alignment in this case. A column appearing in both the upmost and downmost alignments will be present in all optimal alignments between the two sequences considered.

It is possible to modify the algorithm to produce *all* optimal alignments between s and t. We need to keep a stack with the points at which there are options and backtrack to them to explore all possibilities of reaching (0, 0) through the arrows. However, there might be a very large number of optimal alignments. In these cases, it is often advisable to keep some sort of short representation of all or part of these alignments, for instance, the upmost and downmost ones.

Let us now determine the complexity of the algorithms described in this section. The first two do the initialization and consume time $O(m)$ and $O(n)$, respectively. The last two loops are nested and fill the rest of the matrix. The number of operations performed depends essentially on the number of entries that must be computed, that is, the size of the matrix. Thus, we spend time $O(mn)$ in this part and this is the dominant term in the time complexity. The space used is also proportional to the size of the matrix. Hence, the complexity of the basic algorithm is $O(mn)$ both for time and space. If the sequences have the same or nearly the same length, say *n*, we get $O(n^2)$. That is why we say that these algorithms have quadratic complexity.

The construction of the alignment - given the already filled matrix - is done in time *O(len)*, where *len* is the size of the returned alignment, which is $O(m + n)$.

Let us try to understand this with the help of a small case study. We will work it out in stepwise manner as well as using API to implement the algorithm in an automated manner.

Let us start with some simple alignment examples. It is recommended that this be practiced in **IPython** or **Jupyter** notebook. The reader is also requested to install scikit-bio module. This module does not come with the default Anaconda distribution and needs to be installed separately.

In [17]:

```
# We want plots to appear inside the notebook itself. To accomplish this issue the following command:
%pylab inline
from __future__ import division, print_function
# general module import statements
import numpy as np
from IPython.core.display import HTML
from IPython.core import page
page.page = print
from skbio import DNA
Populating the interactive namespace from numpy and matplotlib
```

Lets define two sequences, seq1 and seq2, and develop an approach for aligning them.

In [18]:

```
seq1 = DNA("TGGCCACCTTGGCCTTGGCCATTGTGGGTG")
seq2 = DNA("TGGCCATTGGCCAATTGTGGGTGTC")
```

Please install IAB module available from

http://readiab.org/

Here's how you'd import a function and use it to accomplisg tasks that follow:

In [19]:

```
from iab.algorithms import show_F
```

In [20]:

```
num_rows = len(seq2)
num_cols = len(seq1)
data = np.zeros(shape=(num_rows, num_cols), dtype=np.int)
# the following command displays the alignment of the two sequences in a nice tabular format
HTML(show_F(seq1, seq2, data))
```

Out[20]:

	T	G	G	C	C	A	C	C	T	T	G	G	C	C	T	T	G	G	C	C	A	T	T	G	T	G	G	G	T	G
T	0	0	0	0	0	0	0	0	0	0	0	0	0	0	0	0	0	0	0	0	0	0	0	0	0	0	0	0	0	0
G	0	0	0	0	0	0	0	0	0	0	0	0	0	0	0	0	0	0	0	0	0	0	0	0	0	0	0	0	0	0
G	0	0	0	0	0	0	0	0	0	0	0	0	0	0	0	0	0	0	0	0	0	0	0	0	0	0	0	0	0	0
C	0	0	0	0	0	0	0	0	0	0	0	0	0	0	0	0	0	0	0	0	0	0	0	0	0	0	0	0	0	0
C	0	0	0	0	0	0	0	0	0	0	0	0	0	0	0	0	0	0	0	0	0	0	0	0	0	0	0	0	0	0
A	0	0	0	0	0	0	0	0	0	0	0	0	0	0	0	0	0	0	0	0	0	0	0	0	0	0	0	0	0	0
T	0	0	0	0	0	0	0	0	0	0	0	0	0	0	0	0	0	0	0	0	0	0	0	0	0	0	0	0	0	0
T	0	0	0	0	0	0	0	0	0	0	0	0	0	0	0	0	0	0	0	0	0	0	0	0	0	0	0	0	0	0
G	0	0	0	0	0	0	0	0	0	0	0	0	0	0	0	0	0	0	0	0	0	0	0	0	0	0	0	0	0	0
G	0	0	0	0	0	0	0	0	0	0	0	0	0	0	0	0	0	0	0	0	0	0	0	0	0	0	0	0	0	0
C	0	0	0	0	0	0	0	0	0	0	0	0	0	0	0	0	0	0	0	0	0	0	0	0	0	0	0	0	0	0
C	0	0	0	0	0	0	0	0	0	0	0	0	0	0	0	0	0	0	0	0	0	0	0	0	0	0	0	0	0	0
A	0	0	0	0	0	0	0	0	0	0	0	0	0	0	0	0	0	0	0	0	0	0	0	0	0	0	0	0	0	0
A	0	0	0	0	0	0	0	0	0	0	0	0	0	0	0	0	0	0	0	0	0	0	0	0	0	0	0	0	0	0
T	0	0	0	0	0	0	0	0	0	0	0	0	0	0	0	0	0	0	0	0	0	0	0	0	0	0	0	0	0	0
T	0	0	0	0	0	0	0	0	0	0	0	0	0	0	0	0	0	0	0	0	0	0	0	0	0	0	0	0	0	0
G	0	0	0	0	0	0	0	0	0	0	0	0	0	0	0	0	0	0	0	0	0	0	0	0	0	0	0	0	0	0
T	0	0	0	0	0	0	0	0	0	0	0	0	0	0	0	0	0	0	0	0	0	0	0	0	0	0	0	0	0	0
G	0	0	0	0	0	0	0	0	0	0	0	0	0	0	0	0	0	0	0	0	0	0	0	0	0	0	0	0	0	0
G	0	0	0	0	0	0	0	0	0	0	0	0	0	0	0	0	0	0	0	0	0	0	0	0	0	0	0	0	0	0
G	0	0	0	0	0	0	0	0	0	0	0	0	0	0	0	0	0	0	0	0	0	0	0	0	0	0	0	0	0	0
T	0	0	0	0	0	0	0	0	0	0	0	0	0	0	0	0	0	0	0	0	0	0	0	0	0	0	0	0	0	0
G	0	0	0	0	0	0	0	0	0	0	0	0	0	0	0	0	0	0	0	0	0	0	0	0	0	0	0	0	0	0
T	0	0	0	0	0	0	0	0	0	0	0	0	0	0	0	0	0	0	0	0	0	0	0	0	0	0	0	0	0	0

C	0	0	0	0	0	0	0	0	0	0	0	0	0	0	0	0	0	0	0	0	0	0	0	0	0	0	0	0	0	0

In [21]:

```
for row_number, row_character in enumerate(seq2):
    for col_number, col_character in enumerate(seq1):
        if row_character == col_character:
            data[row_number, col_number] = 1
# The zeros are made invisible and empty spaces appear in place of them.
HTML(show_F(seq1, seq2, data, hide_zeros=True))
```

Out[21]:

	T	G	G	C	C	A	C	C	T	T	G	G	C	C	T	T	G	G	C	C	A	T	T	G	T	G	G	G	T	G
T	1								1	1					1	1						1	1		1				1	
G		1	1								1	1					1	1						1		1	1	1		1
G		1	1								1	1					1	1						1		1	1	1		1
C				1	1		1	1					1	1					1	1										
C				1	1		1	1					1	1					1	1										
A						1															1									
T	1								1	1					1	1						1	1		1				1	
T	1								1	1					1	1						1	1		1				1	
G		1	1								1	1					1	1						1		1	1	1		1
G		1	1								1	1					1	1						1		1	1	1		1
C				1	1		1	1					1	1					1	1										
C				1	1		1	1					1	1					1	1										
A						1															1									
A						1															1									
T	1								1	1					1	1						1	1		1				1	
T	1								1	1					1	1						1	1		1				1	
G		1	1								1	1					1	1						1		1	1	1		1
T	1								1	1					1	1						1	1		1				1	
G		1	1								1	1					1	1						1		1	1	1		1
G		1	1								1	1					1	1						1		1	1	1		1
G		1	1								1	1					1	1						1		1	1	1		1
T	1								1	1					1	1						1	1		1				1	
G		1	1								1	1					1	1						1		1	1	1		1
T	1								1	1					1	1						1	1		1				1	
C				1	1		1	1					1	1					1	1										

Next we'll identify the longest stretches of non-zero characters, which we'll refer to here as the *diagonals*. Diagonals indicate segments of the two sequences that are identical and uninterrupted by mismatched characters (substitution events) or indel events.

We can identify the longest diagonals as follows:

In [22]:

```
# create a copy of our data matrix to work with, so we
# leave the original untouched.
summed_data = data.copy()
# iterate over the cells in our data matrix, starting in
# the second row and second column
for i in range(1, summed_data.shape[0]):
    for j in range(1, summed_data.shape[1]):
        # if the value in the current cell is greater than zero
        # (i.e., the characters at the corresponding pair of
        # sequence positions are the same), add the value from the
        # cell that is diagonally up and to the left.
        if summed_data[i, j] > 0:
            summed_data[i, j] += summed_data[i-1][j-1]


# Identify the longest diagonal
print("The longest diagonal is %d characters long." % summed_data.
max())
HTML(show_F(seq1, seq2, summed_data, hide_zeros=True))
```

Out[22]:

	T	G	G	C	C	A	C	C	T	T	G	G	C	C	T	T	G	G	C	C	A	T	T	G	T	G	G	G	T	G
T	1								1	1					1	1						1	1		1				1	
G		2	1								2	1					2	1						2		2	1	1		2
G		1	3								1	3					1	3						1		1	3	2		1
C				4	1		1	1					4	1					4	1										
C				1	5		1	2					1	5					1	5										
A						6															6									

T	1								1	1					1	1						7	1		1				1	
T	1								1	2					1	2						1	8		1				1	
G		2	1								3	1					3	1						9		2	1	1		2
G		1	3								1	4					1	4						1		1	3	2		1
C				4	1		1	1					5	1					5	1										
C				1	5		1	2					1	6					1	6										
A						6															7									
A						1															1									
T	1								1	1					1	1						2	1		1				1	
T	1								1	2					1	2						1	3		1				1	
G		2	1								3	1					3	1						4		2	1	1		2
T	1								1	1					1	1						1	1		5				2	
G		2	1								2	1					2	1						2		6	1	1		3
G		1	3								1	3					1	3						1		1	7	2		1
G		1	2								1	2					1	2						1		1	2	8		1
T	1								1	1					1	1						1	1		2				9	
G		2	1								2	1					2	1						2		3	1	1		10
T	1								1	1					1	1						1	1		3				2	
C				1	1		1	1					1	1					1	1										

In [23]:

```
from skbio import Protein
seq1 = Protein("HEAGAWGHEE")
seq2 = Protein("PAWHEAE")
```

A pair of sequences can be aligned in different ways. *Needleman-Wunsch* provides the best alignment, as defined by its score. Here we'll compute two new matrices that together allow us to determine the highest alignment score given the sequences and the substitution matrix, and to transcribe the aligned sequences. These matrices are

- the dynamic programming matrix, or ***F*** and,
- the traceback matrix, or ***T***

F and *T* are defined at the same time.

F looks a lot like the matrix we defined in our simplistic example above, but it has one extra row and column, corresponding to the start of each of the sequences (a state that is independent of the first residue of the sequences that is important for our algorithm). *F* keeps track of the best score of the alignment through the corresponding pair of positions, if the alignment were to terminate at that pair of positions.

Because there are multiple possible alignments that a score in *F* can be derived from, we use our second matrix, *T*, to track which single alignment led to each score in *F*. *T* has the same shape (i.e., numbers of rows and columns) as *F*, and its values encode information about how the sequences were aligned to result in the score in the corresponding cell in *F*.

Prior to initialization, *F* and *T* would look like the following.

In [24]:

```
num_rows = len(seq2) + 1
num_cols = len(seq1) + 1
F = np.zeros(shape=(num_rows, num_cols), dtype=np.int)
HTML(show_F(seq1, seq2, F))
```

Out[24]:

		H	E	A	G	A	W	H	E	E
	0	0	0	0	0	0	0	0	0	0
P	0	0	0	0	0	0	0	0	0	0
A	0	0	0	0	0	0	0	0	0	0
W	0	0	0	0	0	0	0	0	0	0
H	0	0	0	0	0	0	0	0	0	0
E	0	0	0	0	0	0	0	0	0	0
A	0	0	0	0	0	0	0	0	0	0
E	0	0	0	0	0	0	0	0	0	0

In [25]:

```
from iab.algorithms import show_T
T = np.full(shape=(num_rows, num_cols), fill_value=" “, dtype=np.str)
HTML(show_T(seq1, seq2, T))
```

Now, let us compute ***F*** and ***T***.

The first row and column of ***F*** are initialized using the following formulas. *d* in these formulas is a value referred to as the **gap penalty**. This is a constant value that is subtracted from the score of the alignment every time a gap character has to be introduced to align the sequences.

We'll use a constant value of d=8 (it's positive because we subtract it) for now, and explore its use more shortly. *i* is the row number in ***F***, and *j* is the column number in ***F***.

$$F(0,0) = 0$$
$$F(i,0) = F(i-1,0) - d$$
$$F(0,j) = F(0,j-1) - d$$

As an exercise, try computing the values for the cells in the first four rows in column zero and the first four columns in row zero of ***F***. What you'll notice is that the score that you compute for most of the cells (all of them except for F(0,0)) depends on the score at another position in ***F***. In a second matrix, ***T***, draw an arrow from the cell that you're currently defining the score for in ***F*** to the cell whose score it depends on. If the score depends on the cell above, you'd draw an up arrow (↑). If the score depends on the cell to the left, you'd draw a left arrow (←). If the score doesn't depend on any other cell (you should have only one of these), indicate that with a bullet (•). Initializing ***F*** would result in the following.

In [26]:

```
d = 8
F[0][0] = 0
for i in range(1, num_rows):
    F[i][0] = F[i-1][0] – d
for j in range(1, num_cols):
    F[0][j] = F[0][j-1] – d
HTML(show_F(seq1, seq2, F))
```

Out[26]:

		H	E	A	G	A	W	G	H	E	E
	0	-8	-16	-24	-32	-40	-48	-56	-64	-72	-80
P	-8	0	0	0	0	0	0	0	0	0	0
A	-16	0	0	0	0	0	0	0	0	0	0
W	-24	0	0	0	0	0	0	0	0	0	0
H	-32	0	0	0	0	0	0	0	0	0	0
E	-40	0	0	0	0	0	0	0	0	0	0
A	-48	0	0	0	0	0	0	0	0	0	0
E	-56	0	0	0	0	0	0	0	0	0	0

```
In [27]:
T[0][0] = "•"
for i in range(1, num_rows):
    T[i][0] = "↑"

for j in range(1, num_cols):
    T[0][j] = "←"

HTML(show_T(seq1, seq2, T))
```

Out[27]:

		H	E	A	G	A	W	G	H	E	E
	•	←	←	←	←	←	←	←	←	←	←
P	↑										
A	↑										
W	↑										
H	↑										
E	↑										
A	↑										
E	↑										

Next, we'll compute the scores for all of the other cells in ***F***, starting at position (1,1). In *Needleman-Wunsch* alignment, the score ***F*** for cell (*i,j*) (when $i > 0$ and $j > 0$) is computed as the maximum of three possible values. *s* refers to the substitution matrix, and c_i and c_j refer to characters in seq1 and seq2.

$$F(i,j) = max \begin{cases} F(i-1, j-1) + s(c_i, c_j) \\ F(i-1, j) - d \\ F(i, j-1) - d \end{cases}$$

Let us speak English now. Describing the scoring function in English, we score a cell with the maximum of three values: either the value of the cell up and to the left plus the score for the substitution taking place in the current cell (which you find by looking up the substitution in the substitution matrix); the value of the cell above minus the gap penalty; or the value of the cell to the left minus the gap penalty. In this way, you're determining whether the best (highest) score is obtained by inserting a gap in sequence 1 (corresponding to F(i−1,j)−d), inserting a gap in sequence 2 (corresponding to F(i,j−1)−d), or aligning the characters in sequence 1 and sequence 2 (corresponding to F(i−1,j−1)+s(ci,cj)).

Local Alignment

A *local alignment* between *s* and *t* is an alignment between a substring of *s* and a substring of *t*.

The above given definition is the simplest possible one. We will now venture into understanding this at a deeper level.

In this section we present an algorithm to find the highest scoring local alignments between two sequences. This algorithm is a variation of the basic algorithm, discussed in the previous section. The main data structure is, as before, an $(m + 1) \times (n + 1)$ array. Only this time the interpretation of the array values is different. Each entry (i, j) will hold the highest score of an alignment between a *suffix* of $s[1..i]$ and a *suffix* of $t[1..j]$. The first row and the first column are initialized with zeros.

For any entry (i, j), there is always the alignment between the empty suffixes of $s[1..i]$ and $t[1..j]$, which has score zero; therefore the array will have all entries greater than or equal to zero. This explains in part the initialization above.

Following initialization, the array can be filled in the usual way, with $a[i, j]$ depending on the value of three previously computed entries. The resulting recurrence is

$$\mathrm{a}[\mathrm{i},\mathrm{j}] = \max \begin{cases} a[i, j-1] + g \\ a[i-1, j-1] + p(i, j) \\ a[i-1, j] + g \\ 0 \end{cases}$$

that is, the same as in the basic algorithm, except that now we have a fourth possibility, not available in the global case, of an empty alignment.

In the end, it suffices to find the maximum entry in the whole array. This will be the score of an optimal local alignment. Any entry containing this value can be used as a starting point to get such an alignment. The rest of the alignment is obtained tracing back as usual, but stopping as soon as we reach an entry with no arrow going out. Alternatively, we can stop as soon as we reach an entry with value zero.

In general, when doing local comparison, we are interested not only in the optimal alignments, but also in near optimal alignments with scores above a certain threshold. References to methods for retrieving near optimal alignments are given at the end of this chapter.

Let us now take up the popular local alignment tool BLAST.

Blast

BLAST is an acronym for **Basic Local Alignment Search Tool**. It is important to observe that here the term *database* refers simply to a usually large set of catalogued sequences. It does not imply any extra capabilities of fast access, data sharing, and so on, commonly found in standard database management systems. For us, therefore, this "database" is merely a collection of sequences, although sequence information is copiously complemented with additional information such as the origin of the data, bibliographic references, sequence function (if known), and others. BLAST returns a list of *high-scoring segment pairs* between the query sequence and sequences in the database. Before explaining what this means and how BLAST obtains its results, we will briefly introduce some terminology on segment pairs to keep this discussion consistent with the original paper describing BLAST.

A *segment* is a substring of a sequence. Given two sequences, a *segment pair* between them is a pair of segments of the same length, one from each sequence. Because the substrings in a segment pair have the same length, we can form a gapless alignment with them. This alignment can be scored using a matrix of substitution scores. No gap-penalty functions are needed, as there are no gaps. The score thus obtained is by definition the score of the segment pair. An example of a segment pair scored with the PAM120 substitution matrix is given below

K	A	L	M	R
V	A	K	N	S
-4	3	-4	-3	-1

This gives a total of -9.

We are now in possession of all the information necessary to describe precisely what BLAST does. Let us divide the process into the following steps for our comprehension:

- Given a query sequence, BLAST returns all segment pairs between the query and a database sequence with scores above a certain threshold *S*.
- The parameter *S* can be set by the user, although a default value is provided in most servers that run the program.
- We have mentioned that a characteristic of reported alignments is the absence of gaps. This is actually a major reason why BLAST is so fast, since looking for good alignments with gaps is a good deal more time consuming.
- A *maximum segment pair* (MSP) between two sequences is a segment pair of maximum score. This score is a measure of sequence similarity and can be computed precisely by dynamic programming. However, BLAST estimates this number much faster than any dynamic programming method.
- The program also returns *locally* maximal segment pairs, that is, those that cannot be improved further by extending or shortening them.

Working of blast algorithm

The BLAST approach to computing high-scoring segment pairs is as follows. It finds certain "seeds," which are very short segment pairs between the query and a database sequence. These seeds are then extended in both directions, without including gaps, until the maximum possible score for extensions of this particular seed is reached. Not all extensions are looked at. The program has a criterion to stop extensions when the score falls below a carefully computed limit. There is a very small chance of the right extension not being found due to this time optimization, but in practice this trade off is highly acceptable.

We can now narrow the steps of BLAST to just three as given below (in simple English):

- Compile a list of high-scoring strings (or words in BLAST jargon)
- Search for hits - each hit gives a seed.
- Extend seeds.

We explain this for protein sequences and then take an example for illustration. For protein sequences, the list of high-scoring words consists of all words with w characters (called tu-mers) that score at least T with some w-mer of the query, using a PAM matrix to compute scores. Here w and T are program parameters. Note that this list may not contain all query w-mers! If a query w-mer consists of very common amino acids, it may be left out because even its score with itself may fall below T. However, there is an option to force inclusion of all query w-mers. The recommended value of w, the seed size, is 4 for protein searches.

Two approaches are tried for scanning the database in search for hits in the list constructed in the previous step. One of them is to arrange the list words into a hash table. Then, for each database word of size w, it is easy to get their index in the table and compare it to the words there, which will be a small fraction of all list words.

The second method uses a *deterministic finite automaton* to search for hits. This device has states and transitions and operates like a machine. It begins in a fixed initial state, and for each character in the database a transition is made to another state. Depending on the state and on the transition, a word from the list is recognized. The automaton is built only once, using the list of high-scoring words as input, and is a compact way of storing all these words. The search is fast, as it requires only a transition per character.

The final extension is straightforward. As we mentioned, to save time the algorithm stops when the score falls a certain distance below the best one obtained to that point for shortest extensions. This is done in both directions, and the high-scoring segment pair originated from this seed is kept. There is a small probability of missing important extensions, but it is negligible.

Extension is based on a well-founded statistical theory that gives exact distribution of gapless local maximum score for random sequences, and permits a very accurate computation of the probability that the segment pair found could be possible due to chance alone. The smaller this probability, the more significant is the match.

We briefly sketch now the main points of this statistical theory. The distribution of the MSP score (defined above) for random sequences *s* and *t* of lengths *m* and *n,* respectively, can be accurately approximated as described next. This approximation gets better as *m* and *n* increase.

Given a matrix of replacement costs s_{ij} for the pairs of characters in the alphabet, and the probability p_i of occurrence of each individual character in the sequences, we first compute a value *l,* solving the equation:

$$\sum_{i,j} p_i p_j e^{s_{ij}} = 1$$

The parameter 1 is the unique positive solution to this equation and can be obtained by Newton's method. Once *k* is known, the expected number of distinct segment pairs between *s* and *t* with score above *S* is

where *K* is a calculable constant. Actually, the distribution of the number of segment pairs scoring above *S* is a Poisson distribution with mean given by the previous formula. From this, it is easy to derive expressions for useful quantities like the average score, intervals where the score will fall 90% of the time, and so on.

With this basic theory in mind we will now take up illustration. Let us delve into some practical situations, wherein you will be using the concepts learnt so far. Actually, we would take an approach where we will do this by examples.

Accessing Databases

We will look at how to access online repositories from your computers. We use **BioPython** for this. You can also look at packages available under **Bioconductor** in **R**, for accomplishing the same.

```
from Bio import Entrez, Medline, SeqIO
Entrez.email = "your_mail_address@here"
#This gives you the list of available databases
handle = Entrez.einfo()
rec = Entrez.read(handle)
```

```
print(rec)
{'DbList': ['pubmed', 'protein', 'nuccore', 'nucleotide', 'nucgss', 'nucest', 'structure', 'sparcle', 'genome', 'annotinfo', 'assembly', 'bioproject', 'biosample', 'blastdbinfo', 'books', 'cdd', 'clinvar', 'clone', 'gap', 'gapplus', 'grasp', 'dbvar', 'gene', 'gds', 'geoprofiles', 'homologene', 'medgen', 'mesh', 'ncbisearch', 'nlmcatalog', 'omim', 'orgtrack', 'pmc', 'popset', 'probe', 'proteinclusters', 'pcassay', 'biosystems', 'pccompound', 'pcsubstance', 'pubmedhealth', 'seqannot', 'snp', 'sra', 'taxonomy', 'biocollections', 'unigene', 'gencoll', 'gtr']}
from Bio.Blast import NCBIWWW
result_handle = NCBIWWW.qblast("blastn", "nt", "8332116")
```

Now that we've got a handle, we are ready to parse the output. The code to parse it is really quite small. If you expect a single BLAST result (i.e., you used a single query):

```
from Bio.Blast import NCBIXML
blast_record = NCBIXML.read(result_handle)
```

Please refer to **BioPython** documentation for parsing BLAST output.

Let us take another illustration. We are in search of gene coding for *insulin* in homo sapiens. This can be done by specifying the Gene and organism name in the search call to the remote online repository.

```
handle = Entrez.esearch(db="nucleotide", term='INS[Gene Name] AND "Homo sapiens"[Organism]')
rec_list = Entrez.read(handle)
if rec_list['RetMax'] < rec_list['Count']:
    handle = Entrez.esearch(db="nucleotide", term='INS[Gene Name] AND "Homo sapiens"[Organism]',
                            retmax=rec_list['Count'])
    rec_list = Entrez.read(handle)
id_list = rec_list['IdList']
hdl = Entrez.efetch(db='nucleotide', id=id_list, rettype='gb')
# parsing each record found in the above search
recs = list(SeqIO.parse(hdl, 'gb'))
# print the records (optional); still a good practice to check the results
```

```
recs
[SeqRecord(seq=Seq('AGCCCTCCAGGACAGGCTGCATCAGAAGAGGCCATCAAGCAGGT-CTGTTCCAAG...AAA', IUPACAmbiguousDNA()), id='NM_001291897.1', name='NM_001291897', description='Homo sapiens insulin (INS), transcript variant 4, mRNA.', dbxrefs=[]),
SeqRecord(seq=Seq('AGCCCTCCAGGACAGGCTGCATCAGAAGAGGCCATCAAGCAGGTCTG-TTCCAAG...AAA', IUPACAmbiguousDNA()), id='NM_001185098.1', name='NM_001185098', description='Homo sapiens insulin (INS), transcript variant 3, mRNA.', dbxrefs=[]),
SeqRecord(seq=Seq('AGCCCTCCAGGACAGGCTGCATCAGAAGAGGCCATCAAGCAGGTCTGT-TCCAAG...AAA', IUPACAmbiguousDNA()), id='NM_001185097.1', name='NM_001185097', description='Homo sapiens insulin (INS), transcript variant 2, mRNA.', dbxrefs=[]),
SeqRecord(seq=Seq('AGCCCTCCAGGACAGGCTGCATCAGAAGAGGCCATCAAGCAGATCACT-GTCCTT...AAA', IUPACAmbiguousDNA()), id='NM_000207.2', name='NM_000207', description='Homo sapiens insulin (INS), transcript variant 1, mRNA.', dbxrefs=[]),
SeqRecord(seq=Seq('GAGGTGCGGATCCTGGGCGGCCAGGGAAGGTCTCTGCCGCCAGGGAAG-TGTCCC...CTC', IUPACAmbiguousDNA()), id='NG_050578.1', name='NG_050578', description='Homo sapiens INS-IGF2 readthrough (INS-IGF2), RefSeqGene on chromosome 11.', dbxrefs=[]),
SeqRecord(seq=Seq('GGCGGCCAGGGAAGGTCTCTGCCGCCAGGGAAGTGTCCCAGAGACCCC-TGGAGG...GCC', IUPACAmbiguousDNA()), id='NG_007114.1', name='NG_007114', description='Homo sapiens insulin (INS), RefSeqGene on chromosome 11.', dbxrefs=[]),
SeqRecord(seq=Seq('CTGGGGACCTGACCCAGCCGCAGCCTTTGTGAACCAACACCTGTGCGG-CTCACA...CTA', IUPACAmbiguousDNA()), id='JF909299.1', name='JF909299', description='Homo sapiens insulin (INS) mRNA, partial cds.', dbxrefs=[]),
```

SeqRecord(seq=Seq('GTACAAAAAAGCAGGCTCCACCA TGGCCCTGTGGATGCGCCTCCTGCC-CCTGCT...TAC', IUPACAmbiguousDNA()), id='DQ896283.2', name='DQ896283', description='SyntheticconstructHomosapienscloneIMAGE:100010743; FLH192918.01L; RZPDo839A1068D insulin (INS) gene, encodes complete protein.', dbxrefs=[]),

SeqRecord(seq=Seq('CTCGAGGGGCCTAGACATTGCCC TCCAGAGAGAGCACCCAACACCCTCC-AGGCT...TCA', IUPACAmbiguousDNA()), id='AH002844.2', name='AH002844', description='Homo sapiens insulin (INS) gene, complete cds.', dbxrefs=[]),

SeqRecord(seq=Seq('GAGCAGGGCTGGACCTGTGAGCC CAGGTCACAGATGAGAAACCGAC-CCCTGGT...AAG', IUPACAmbiguousDNA()), id='AH012037.2', name='AH012037', description='Homo sapiens haplotype ICa tyrosine hydroxylase (TH) gene, partial sequence; insulin (INS) gene, complete cds; and insulin-like growth factor 2 (IGF2) gene, partial sequence.', dbxrefs=[]),

SeqRecord(seq=UnknownSeq(135086622, alphabet = IUPACAmbiguousDNA(), character = 'N'), id='NC_000011.10', name='NC_000011', description='Homo sapiens chromosome 11, GRCh38.p7 Primary Assembly.', dbxrefs=['BioProject:PRJNA168', 'Assembly:GCF_000001405.33']),

SeqRecord(seq=UnknownSeq(134889443, alphabet = IUPACAmbiguousDNA(), character = 'N'), id='NC_018922.2', name='NC_018922', description='Homo sapiens chromosome 11, alternate assembly CHM1_1.1, whole genome shotgun sequence.', db xrefs=['BioProject:PRJNA178030', 'BioSample:SAMN02205338', 'Assembly:GCF_000306695.2']),

SeqRecord(seq=Seq('GTTCGTTGCAACAAATTGATGAGCAATGC TTTTTTATAATGCCAACTTTGTACA...AAC', IUPACAmbiguous-DNA()), id='KR710185.1', name='KR710185', description='Synthetic construct Homo sapiens clone CCSBHm_00010262 INS (INS) mRNA, encodes complete protein.', dbxrefs=[]),

SeqRecord(seq=Seq('GTTCGTTGCAACAAATTGATGAGCAATGC TTTTTTATAATGCCAACTTTGTACA...AAC', IUPACAmbiguous-DNA()), id='KR710184.1', name='KR710184', description='Synthetic construct Homo sapiens clone CCSBHm_00010257 INS (INS) mRNA, encodes complete protein.', dbxrefs=[]),

SeqRecord(seq=Seq('GTTCGTTGCAACAAATTGATGAGCAATGC TTTTTTATAATGCCAACTTTGTACA...AAC', IUPACAmbiguous-DNA()), id='KR710183.1', name='KR710183', description='Synthetic construct Homo sapiens clone CCSBHm_00010256 INS (INS) mRNA, encodes complete protein.', dbxrefs=[]),

SeqRecord(seq=Seq('GTTCGTTGCAACAAATTGATGAGCAATGC TTTTTTATAATGCCAACTTTGTACA...AAC', IUPACAmbiguous-DNA()), id='KR710182.1', name='KR710182', description='Synthetic construct Homo sapiens clone CCSBHm_00010255 INS (INS) mRNA, encodes complete protein.', dbxrefs=[]),

SeqRecord(seq=UnknownSeq(132164248, alphabet = IUPA-CAmbiguousDNA(), character = 'N'), id='CM000262.1', name='CM000262', description='Homo sapiens chromosome 11, whole genome shotgun sequence.', dbxrefs=['BioProject:PRJNA1431', 'BioSample:SAMN02981219']),

SeqRecord(seq=UnknownSeq(3852046, alphabet = IUPACAmbigu-ousDNA(), character = 'N'), id='CH471158.1', name='CH471158', description='Homo sapiens 211000035842702 genomic scaffold, whole genome shotgun sequence.', dbxrefs=['BioProject:PRJNA1431', 'BioSample:SAMN02981219']),

SeqRecord(seq=Seq('AGCCCTCCAGGACAGGCTGCATCAGAAG AGGCCATCAAGCAGATCACTGTCCTT...AAA', IUPACAmbigu-ousDNA()), id='BC005255.1', name='BC005255', description='Homo sapiens insulin, mRNA (cDNA clone MGC:12292 IMAGE:3950204), complete cds.', dbxrefs=[]),

SeqRecord(seq=UnknownSeq(606048, alphabet = IUPACAmbigu-ousDNA(), character = 'N'), id='NT_086367.4', name='NT_086367', description='Homo sapiens chromosome 11 sequence, ENCODE region ENm011.', dbxrefs=[]),

SeqRecord(seq=Seq('GCTGCATCAGAAGAGGCCATCAAGCACAT CACTGTCCTTCTGCCATGGCCCTGT...ATG', IUPACAmbiguousD-NA()), id='JQ951950.1', name='JQ951950', description='Homo sapiens preproinsulin (INS) mRNA, complete cds.', dbxrefs=[]),

SeqRecord(seq=Seq('CTGCCAGCACTGCCTGTCTGAGGAGCCTG AGATTCGGGCCTTCGACCCTGAGGC...TCA', IUPACAmbiguousDNA()), id='L15440.1', name='HUMINSTHIG', description="Homo sapiens tyrosine hydroxylase (TH) gene, 3' end; insulin (INS) gene, complete cds; insulin-like growth factor 2 (IGF2) gene, 5' end.", dbxrefs=[]),

SeqRecord(seq=Seq('GCGATCGCCATGGCCCTGTGGATGCGCCT CCTGCCCCTGCTGGCGCTGCTGGCC...AAC', IUPACAmbiguousDNA()), id='AB587580.1', name='AB587580', description='Synthetic construct DNA, clone: pF1KB8864, Homo sapiens INS gene for insulin, without stop codon, in Flexi system.', dbxrefs=[]),

SeqRecord(seq=Seq('CTGGGGCTGCTGTCCTAAGGCAGGGTGG GAACTAGGCAGCCAGCAGGGAGGGGA...CCC', IUPACAmbiguousDNA()), id='M10039.1', name='HUMINSPR', description="Human alpha-type insulin gene and 5' flanking polymorphic region.", dbxrefs=[]),

SeqRecord(seq=Seq('TAGGGGCTCAGCCCAGATGACACTATGGG GGTGATGGTGTCATGGGACCTGGCC...CTC', IUPACAmbiguousDNA()), id='J00267.1', name='HUMINS1UB', description="Human insulin gene: beta allele 5'flank (ullrich).", dbxrefs=[]),

SeqRecord(seq=Seq('TAGGGGCTCAGCCCAGATGACACTATGGG GGTGATGGTGTCATGGGACCTGGCC...CTC', IUPACAmbiguousDNA()), id='J00266.1', name='HUMINS1UA', description="Human insulin gene: alpha allele 5'flank (ullrich).", dbxrefs=[]),

SeqRecord(seq=Seq('GTACAAAAAGCAGGCTCCACCATGGCC CTGTGGATGCGCCTCCTGCCCCTGCT...TAC', IUPACAmbiguousDNA()), id='DQ893040.2', name='DQ893040', description='Synthetic construct clone IMAGE:100005670; FLH192922.01X; RZPDo839A1078D insulin (INS) gene, encodes complete protein.', dbxrefs=[]),

SeqRecord(seq=Seq('AGCAGGTCTGTTCCAAGGGCCTTTGCGTC AGGTGGGCTCAGGGTTCCAGGGTGG...AGC', IUPACAmbiguousDNA()), id='AJ009655.1', name='AJ009655', description='Homo sapiens ins gene, partial.', dbxrefs=[])]

```
# you can pull out a record of interest from the above list; typically we do this if we are interested in a specific gene identifier

for rec in recs:

    if rec.name == 'NM_000207':

        break

print(rec.name)                    # print record name

print(rec.description)  # print description of the gene

NM_000207

Homo sapiens insulin (INS), transcript variant 1, mRNA.

# print different features of the gene(s)

for feature in rec.features:

    if feature.type == 'gene':

        print(feature.qualifiers['gene'])

    elif feature.type == 'exon':

        loc = feature.location

        print('Exon', loc.start, loc.end, loc.strand)

    else:

        print('not processed:\n%s' % feature)

not processed:

type: source

location: [0:469](+)

qualifiers:

    Key: chromosome, Value: ['11']

    Key: db_xref, Value: ['taxon:9606']

    Key: map, Value: ['11p15.5']
```

Key: mol_type, Value: ['mRNA']

Key: organism, Value: ['Homo sapiens']

['INS']

Exon 0 42 1

not processed:

type: STS

location: [15:186](+)

qualifiers:

Key: db_xref, Value: ['UniSTS:155248']

Key: gene, Value: ['INS']

Key: gene_synonym, Value: ['IDDM; IDDM1; IDDM2; ILPR; IRDN; MODY10']

Key: standard_name, Value: ['GDB:181496']

Exon 42 246 1

not processed:

type: CDS

location: [59:392](+)

qualifiers:

Key: codon_start, Value: ['1']

Key: db_xref, Value: ['CCDS:CCDS7729.1', 'GeneID:3630', 'HGNC:HGNC:6081', 'MIM:176730']

Key: gene, Value: ['INS']

Key: gene_synonym, Value: ['IDDM; IDDM1; IDDM2; ILPR;

IRDN; MODY10']

Key: note, Value: ['proinsulin; preproinsulin']

Key: product, Value: ['insulin preproprotein']

Key: protein_id, Value: ['NP_000198.1']

Key: translation, Value: ['MALWMRLLPLLALLALWGPDPAAAF-VNQHLCGSHLVEALYLVCGERGFFYTPKTRREAEDLQVGQVEL-GGGPGAGSLQPLALEGSLQKRGIVEQCCTSICSLYQLENYCN']

not processed:

type: sig_peptide

location: [59:131](+)

qualifiers:

Key: gene, Value: ['INS']

Key: gene_synonym, Value: ['IDDM; IDDM1; IDDM2; ILPR; IRDN; MODY10']

Key: inference, Value: ['COORDINATES: ab initio prediction:SignalP:4.0']

not processed:

type: proprotein

location: [131:389](+)

qualifiers:

Key: gene, Value: ['INS']

Key: gene_synonym, Value: ['IDDM; IDDM1; IDDM2; ILPR; IRDN; MODY10']

Key: product, Value: ['proinsulin']

not processed:

type: mat_peptide

location: [131:221](+)

qualifiers:

Key: gene, Value: ['INS']

Key: gene_synonym, Value: ['IDDM; IDDM1; IDDM2; ILPR; IRDN; MODY10']

Key: product, Value: ['insulin B chain']

not processed:

type: mat_peptide

location: [227:320](+)

qualifiers:

Key: gene, Value: ['INS']

Key: gene_synonym, Value: ['IDDM; IDDM1; IDDM2; ILPR; IRDN; MODY10']

Key: product, Value: ['C-peptide']

not processed:

type: mat_peptide

location: [326:389](+)

qualifiers:

Key: gene, Value: ['INS']

Key: gene_synonym, Value: ['IDDM; IDDM1; IDDM2; ILPR; IRDN; MODY10']

Key: product, Value: ['insulin A chain']

not processed:

type: STS

location: [59:391](+)

qualifiers:

Key: db_xref, Value: ['UniSTS:270424']

Key: gene, Value: ['INS']

Key: gene_synonym, Value: ['IDDM; IDDM1; IDDM2; ILPR; IRDN; MODY10']

Key: standard_name, Value: ['PMC123023P3']

not processed:

type: STS

location: [67:368](+)

qualifiers:

Key: db_xref, Value: ['UniSTS:272021']

Key: gene, Value: ['INS']

Key: gene_synonym, Value: ['IDDM; IDDM1; IDDM2; ILPR; IRDN; MODY10']

Key: standard_name, Value: ['PMC21231P1']

not processed:

type: STS

location: [71:371](+)

qualifiers:

Key: db_xref, Value: ['UniSTS:259556']

Key: gene, Value: ['INS']

Key: gene_synonym, Value: ['IDDM; IDDM1; IDDM2; ILPR; IRDN; MODY10']

Key: standard_name, Value: ['Ins2']

not processed:

type: STS

location: [78:390](+)

qualifiers:

Key: db_xref, Value: ['UniSTS:256685']

Key: gene, Value: ['INS']

Key: gene_synonym, Value: ['IDDM; IDDM1; IDDM2; ILPR; IRDN; MODY10']

Key: standard_name, Value: ['Ins1']

not processed:

type: STS

location: [197:385](+)

qualifiers:

Key: db_xref, Value: ['UniSTS:265494']

Key: gene, Value: ['INS']

Key: gene_synonym, Value: ['IDDM; IDDM1; IDDM2; ILPR; IRDN; MODY10']

Key: standard_name, Value: ['PMC24644P6']

Exon 246 465 1

not processed:

type: STS

location: [246:465](+)

qualifiers:

Key: db_xref, Value: ['UniSTS:155046']

Key: gene, Value: ['INS']

Key: gene_synonym, Value: ['IDDM; IDDM1; IDDM2; ILPR; IRDN; MODY10']

Key: standard_name, Value: ['GDB:179433']

not processed:

type: regulatory

location: [445:451](+)

qualifiers:

Key: gene, Value: ['INS']

Key: gene_synonym, Value: ['IDDM; IDDM1; IDDM2; ILPR; IRDN; MODY10']

Key: regulatory_class, Value: ['polyA_signal_sequence']

not processed:

type: polyA_site

location: [464:465](+)

qualifiers:

Key: gene, Value: ['INS']

Key: gene_synonym, Value: ['IDDM; IDDM1; IDDM2; ILPR; IRDN; MODY10']

for name, value in rec.annotations.items():

print('%s=%s' % (name, value))

comment=REVIEWED REFSEQ: This record has been curated by NCBI staff. The reference sequence was derived from BC005255.1 and BM510748.1. This sequence is a reference standard in the RefSeqGene project.

On Jun 15, 2006 this sequence version replaced gi:4557670.

Summary: After removal of the precursor signal peptide, proinsulin is post-translationally cleaved into three peptides: the B chain and A chain peptides, which are covalently linked via two disulfide bonds to form insulin, and C-peptide. Binding of insulin to the insulin receptor (INSR) stimulates glucose uptake. A multitude of

mutant alleles with phenotypic effects have been identified. There is a read-through gene, INS-IGF2, which overlaps with this gene at the 5' region and with the IGF2 gene at the 3' region. Alternative splicing results in multiple transcript variants. [provided by RefSeq, Jun 2010].

Transcript Variant: This variant (1) represents the shortest variant. All variants encode the same protein.

Publication Note: This RefSeq record includes a subset of the publications that are available for this gene. Please see the Gene record to access additional publications. SAMEA2153347 [ECO:0000348]

COMPLETENESS: complete on the 3' end.

source=Homo sapiens (human)

taxonomy=['Eukaryota', 'Metazoa', 'Chordata', 'Craniata', 'Vertebrata', 'Euteleostomi', 'Mammalia', 'Eutheria', 'Euarchontoglires', 'Primates', 'Haplorrhini', 'Catarrhini', 'Hominidae', 'Homo']

structured_comment=defaultdict(<type 'dict'>, {'Evidence-Data': {'Transcript exon combination': 'BC005255.1, BU075674.1 [ECO:0000332]', 'RNAseq introns': 'single sample supports all introns'}})

keywords=['RefSeq']

references=[Reference(title='Elaboration of a finite element model of pancreatic islet dielectric response to gap junction expression and insulin release', ...), Reference(title='Upregulation of INS-IGF2 read-through expression and identification of a novel INS-IGF2 splice variant in insulinomas', ...), Reference(title='The influence of INS VNTR class III allele on auxological parameters, glucose, insulin, lipids, and adipocytokines secretion in prepubertal children born small for gestational age', ...), Reference(title="Allele-specific recognition of the 3' splice site of INS intron 1", ...), Reference(title='Global haplotype diversity in the human insulin gene region', ...), Reference(title='Permanent Neonatal Diabetes Mellitus', ...), Reference(title='Nucleotide sequence of a cDNA clone encoding human preproinsulin', ...), Reference(title='A structurally abnormal insulin causing human diabetes', ...), Reference(title='Effects of altered thyroid function on galactosyl diacylglycerol metabolism in myelinating rat brain', ...), Reference(title='Chemistry of insulin; determination of the structure of insulin opens the way to greater understanding of life processes', ...)]

accessions=['NM_000207']

data_file_division=PRI

date=30-APR-2017

organism=Homo sapiens

sequence_version=2

COMPARING MULTIPLE ALIGNMENTS

So far in this chapter we have concentrated on the comparison between a pair of sequences. However we often are given several sequences that we have to align simultaneously in the best possible way. This happens, for example, when we have the sequences for certain proteins that have similar function in a number of different species. We may want to know which parts of these sequences are similar and which parts are different. To get this information, we need to build a **multiple alignment** for these sequences, and that is the topic of this section.

The notion of multiple alignment is a natural generalization of the two-sequence case. Let $s_1, \ldots, s_k$ be a set of sequences over the same alphabet. A multiple alignment involving $s_1, \ldots, s_k$ is obtained by inserting spaces in the sequences in such a way as to make them all of the same size. It is customary to place the extended sequences in a vertical list so that characters - or spaces - in corresponding positions occupy the same column. We further require that no column be made exclusively of spaces. Below is shown a multiple alignment involving four short amino acid sequences. (Multiple alignments are more common with proteins, so in some of our examples in this section use sequences of amino acids.)

One important issue to decide in multiple alignment is the precise definition of the *quality* of an alignment. We will next study one way of scoring a multiple alignment based on pairwise alignments. In addition, scientists also look at multiple alignments by placing the sequences in a tree structure rather than piling them up. This leads to different measures of quality that we discuss in subsequent sections.

MQPILLL

MLR-LL-

MK-ILLL

MPPVLIL

The SP Measure

Scoring a multiple alignment is more complex than its pairwise counterpart. We restrict ourselves to purely additive functions here; that

is, the alignment score is the sum of column scores. Therefore we need a way to assign a score to each column and then add them up to get the alignment score. However, to score a column, we want a function with k arguments, where k is the number of sequences. Each one of these arguments is a character or a space. One way to do that would be to have a fc-dimensional array that could be indexed with the arguments and return the value. The problem with this approach is that it is necessary to specify a value for each possible combination of arguments, and there could be as many as $2^k — 1$ such combinations. A typical value for /c is 10, and that results in more than 1000 possibilities. Some more manageable methods to define such a function are necessary.

Such methods can be obtained by determining "reasonable" properties that such a function should have. First, the function must be independent of the order of arguments. For instance, if a column has I, -, I, V and another has V, I, I, -, they should both receive the same score. Second, it should be a function that rewards the presence of many equal or strongly related residues and penalizes unrelated residues and spaces. A solution that satisfies these properties is the so-called **sum-of-pairs** (SP) function. It is defined as the sum of pairwise scores of all pairs of symbols in the column. For instance, the score of a column with the above content is

$$SP\text{-}score(\mathrm{I}, -, \mathrm{I}, \mathrm{V}) = p(\mathrm{I}, -) + p(\mathrm{I}, \mathrm{I}) + p(\mathrm{I}, \mathrm{V}) + p(-, \mathrm{I}) + p(-, \mathrm{V}) + p(\mathrm{I}, \mathrm{V})$$

where $p(a, b)$ is the pairwise score of symbols a and b. Notice that this may include a space penalty specification when either a or b is a space. This is a very convenient scheme, because it relies on pairwise scores like the ones we use for two-sequence comparison. The SP scoring system is widely used due to its simplicity and effectiveness.

A small but important detail needs to be addressed to complete the definition. Although no column can be composed exclusively by spaces, it is possible to have two or more spaces in a given column. When computing the SP score, we need a value for $p(-, -)$. This value is not specified in two-sequence comparison because it never appears there. However, it is necessary for SP-based multiple comparison. The common practice is to set $p(-, -) = 0$. This may seem strange, given that spaces are generally penalized (i.e., when there is a space, the pairwise score is negative), so two spaces should be even more so. Nevertheless, there are good reasons to define $p(-, -)$ as zero. One of them is related to pairwise alignments again. We often draw conclusions about a multiple alignment by looking at the pairwise alignments it induces. Indeed, in any multiple alignment, we may select two of the sequences and just look at the way they are aligned to each other, forgetting about all the rest. It is not difficult to see that this produces a pairwise alignment, except for the fact that we may have columns with two spaces in them. But then we just remove these columns and derive a true pairwise alignment. This is what is called as the *induced* pairwise alignment, or the projection of a multiple alignment α:

$$SP\text{-}score(\alpha) = \sum_{i<j} \text{score}(\alpha_{ij})$$

where α_{ij} is the pairwise alignment induced by α on sequences s_i and s_j. This is true because it reflects two ways of doing the same thing. We may compute the score of each column and then add all column scores, or compute the score for each induced pairwise alignment and then add these scores. In any case, we are adding, for each column c and for each pair (i, j), the score $p(s'_i[c], s'_j[c])$, where s' indicates the extended sequence (with spaces inserted). But this is true only if $p(-, -) = 0$, because these quantities appear in the first computation only.

Multiple alignment

1 PEAALYGRFT---ILSDVW

2 PEAALYGRFT---IKSDVW

3 PESLAYNKP---SIKSDVW

4 PEALNYGRY---SSSESDVW

5 PEALNYGWY---SSESDVW

6 PEVIRMQDDNPFSFQSDVY

Get only sequences 2 and 4

2 PEAALYGRFT---IKSDVW

4 PEALNYGRY---SSSESDVW

Remove columns with two spaces

PEAALYGRFT-IKSDVW

PEALNYGRY-SSESDVW

Dynamic Programming

"Dynamic programming" is a phrase that appears again and again in computational biology. This is an old name for a technique that appeared in the 1950s, before computer programming was an everyday term, ***so*** do not be fooled by the word "programming" here. Dynamic programming (DP) is an optimization technique: most commonly, it involves finding the optimal solution to a search problem. Many different algorithms have been called (accurately) dynamic programming algorithms, and quite a few important ideas in computational biology fall under this rubric. All these methods have a few basic principles in common, which we will introduce here. But the first thing to do, if you are not already familiar with dynamic programming (DP), is to ignore the name - it is misleading, because it has nothing to do with programming as we know it today.

The fundamental principle that all DP algorithms share is that they first produce optimal solutions to small pieces of the problem, and then glue them together to get a complete, and still optimal, solution. It is not always possible to break a problem up this way; sometimes gluing solutions together is too costly, and sometimes finding optimal solutions to sub-problems is just as hard as solving the whole problem. When a problem does have this property, though, it is said to satisfy the "principle of optimality."

An example will illustrate what we mean by this principle. Suppose we want to find the shortest path between two cities, **A** and B. We can represent shortest path problems as a graph, where each city is a node and edges between nodes are marked with the distances between cities. For all pairs of cities that are directly connected, we can keep a lookup table that contains their distances. We could also use cost instead of distance, in which case we had be finding the cheapest way to get from one city to another.

Now we use the principle of optimality as follows. The idea is to break up the problem into smaller pieces, solve the pieces, and then glue them together. With our path problem, we can break the problem into pieces by first finding the shortest path from A to X, then finding the shortest path from X to B. The length of the path A-X-B is simply the sum of the costs of the two sub-paths. Thus the "gluing" operation is very simple to perform.

But what if the shortest path from A to B does not pass through X at all? This is certainly possible, and if it happens then our algorithm will fail to give us the correct answer. We can solve this by running our algorithm for *all* cities in the graph. In other words, we will compute the shortest path from A to B that passes through every possible city. The shortest of these has to be the shortest overall.

Now we have two sub-problems to solve: find the shortest path from A to X, and find the shortest path from X to B. To solve the sub-problem, we simply call our dynamic programming algorithm itself! This recursive (self-invoking) call is also characteristic of dynamic programming: DP is almost always expressed recursively. So to find the shortest path from A to X, we will find the shortest path from A to *Y*, and from *Y* to X, and glue those together. We again repeat this for all values of Y, this time excluding B. (Note that by excluding B, we make the problem a little smaller.) As we continue making recursive calls, we eventually get to cities that are directly connected to each other. At this point we just look up the answer in our table.

For an example from computational biology, consider the problem of finding genes in DNA. In gene finding, the task is to produce the "optimal" interpretation of a DNA sequence by parsing it into a series of sub-sequences. Each sub-sequence is labeled according to its role in the genome; for example, intergenic **DNA,** exon, intron, and other labels. Each algorithm has its own way of scoring these sub-sequences, and the role of the dynamic program is to guarantee that, whatever the scoring method, the algorithm will find the optimal score. Of course, if the method does not produce accurate scores, then the overall output might be poor.

The dynamic programming version of gene finding says that to score or "parse" a sequence $S(1, n) = s1, s2, \ldots, s_n$, one can first find the optimal score for a sequence $S(1, i) = s_1, s_2, \ldots, s_i$. (The notation $S(i,j)$ indicates the sub-sequence of S beginning with the base at position i and ending with j). If this prefix is scored optimally, then you can "glue" it together with an optimal parse for the rest of the sequence $s_{i+1}, s_{i+2}, \ldots, s_n$. With this formulation, the algorithm can proceed in a *left-to-right* manner through any sequence, adding one base at a time.

With each new base *i,* the algorithm computes the optimal score for the sequence up to *i.* Under the dynamic programming assumption, the optimal score for $S(1, i)$ must be the sum of the scores for $S(1, \boldsymbol{k})$ and $\boldsymbol{S(k} + 1, \text{i})$ for some $\boldsymbol{k}$ between 1 and i. Since we have already computed the optimal scores for all such $\boldsymbol{k}$, we can find the optimal score for i itself by computing all the intermediate sums and choosing the maximum.

Now continuing from the previous section, having decided on a *measure* or *score*, for determining the **quality** of a multiple alignment, we would like to compute the alignments of maximum score, given a set of sequences. These will be our *optimal alignments*.

It is possible to use a dynamic programming approach here, as we did in the two sequence case. Suppose, for simplicity, that we have k sequences, all of the same length n.

We use a k-dimensional array a of length $n + 1$ in each dimension to hold the optimal scores for multiple alignments of prefixes of the sequences. Thus, $a[i_1,..., i_k]$ holds the score of the optimal alignment involving $s_1[1..i_1], ..., s_k[1..i_k]$.

After initializing with $a[0, . . . , 0] \leftarrow 0$, we must fill in this entire array. Just to store it requires $O(n^k)$ space. This is also a lower bound for the computation time, because we have to compute the value of each entry. The actual time complexity is higher for a number of reasons. First, each entry depends on 2^k - 1 previously computed entries, one for each possible composition of the current column of the alignment. In this composition, each sequence can participate with either a character or a space. Because we have k sequences, we have 2^k compositions. Removing the forbidden composition of all spaces, we obtain the final count of 2^k - 1. This incorporates a factor of 2^k to the already exponential time complexity.

Another issue is the computation of column scores. The SP method requires $O(k^2)$steps per column, as there are $k(k — 1)/2$ pairwise scores to add up. Simpler schemes - for instance, just count the umber of nonspace symbols - require at least $O(k)$ steps, given that we have to look at all arguments.

Finally, there is the actual computing of the value of $a[i_1, ..., i_k]$, which involves a maximum operation. Using boldface letters to indicate k-tuples, the command we must perform can be written as

$$a[i] \leftarrow \max_{b \neq 0} \left\{ a[i-b] + SP-score\left(Column(s, i, b)\right) \right\}$$

where b ranges over all non-zero binary vectors of k elements, and

Column(s, i, b) =

with

$$c_j = \begin{cases} s_j[i_j] & if\, b_j = 1 \\ - & if\, b_j = 0. \end{cases}$$

Now let us turn our attention on how to achieve this using computational methods and programming languages. We use BioPython again. There are lots of *algorithms* out there for aligning sequences, both pairwise alignments and multiple sequence alignments. These calculations are relatively slow, and you generally wouldn't want to write such an algorithm in *Python*. Instead, you can use *Biopython* to invoke a command line tool on your behalf. Normally you would:

- Prepare an input file of your unaligned sequences, typically this will be a FASTA file which you might create using Bio. SeqIO.
- Call the command line tool to process this input file, typically via one of Biopython's command line wrappers.
- Read the output from the tool, i.e. your aligned sequences, typically using Bio.AlignIO.

All the command line wrappers we're going to talk about in this chapter follow the same style. You create a command line object specifying the options (e.g. the input filename and the output filename), then invoke this command line via a Python operating system call (e.g. using the *subprocess* module). Most of these wrappers are defined in the Bio.Align.Applications module:

```
import Bio.Align.Applications
dir(Bio.Align.Applications)
['ClustalOmegaCommandline',
 'ClustalwCommandline',
 'DialignCommandline',
 'MSAProbsCommandline',
```

```
'MafftCommandline',
'MuscleCommandline',
'PrankCommandline',
'ProbconsCommandline',
'TCoffeeCommandline',
'_ClustalOmega',
'_Clustalw',
'_Dialign',
'_MSAProbs',
'_Mafft',
'_Muscle',
'_Prank',
'_Probcons',
'_TCoffee',
'__all__',
'__builtins__',
'__doc__',
'__file__',
'__name__',
'__package__',
'__path__']
```

We won't explore all these alignment tools here in the section, just a sample, but the same principles apply.

ClustalW

ClustalW is a popular command line tool for multiple sequence alignment (there is also a graphical interface called ClustalX). Biopython's *Bio. Align.Applications* module has a wrapper for this alignment tool (and several others).

Before trying to use ClustalW from within Python, you should first try running the ClustalW tool yourself by hand at the command line, to familiarise yourself the other options. You'll find the Biopython wrapper is very faithful to the actual command line API:

```
from Bio.Align.Applications import ClustalwCommandline

help(ClustalwCommandline)
Help on class ClustalwCommandline in module Bio.Align.Applications._Clustalw:


class ClustalwCommandline(Bio.Application.AbstractCommandline)
 | Command line wrapper for clustalw (version one or two).
 |
 | http://www.clustal.org/
 |
 | Example:
 | --------
 |
 | >>> from Bio.Align.Applications import ClustalwCommandline
 | >>> in_file = "unaligned.fasta"
 | >>> clustalw_cline = ClustalwCommandline("clustalw2", infile=in_file)
 | >>> print(clustalw_cline)
 | clustalw2 -infile=unaligned.fasta
 |
 | You would typically run the command line with clustalw_cline() or via
 | the Python subprocess module, as described in the Biopython tutorial.
 |
 | Citation:
 | ---------
 |
 | Larkin MA, Blackshields G, Brown NP, Chenna R, McGettigan PA,
 | McWilliam H, Valentin F, Wallace IM, Wilm A, Lopez R, Thompson
```

```
JD,
 |  Gibson TJ, Higgins DG. (2007). Clustal W and Clustal X version 2.0.
 |  Bioinformatics, 23, 2947-2948.
 |
 |  Last checked against versions: 1.83 and 2.1
 |
 |  Method resolution order:
 |      ClustalwCommandline
 |      Bio.Application.AbstractCommandline
 |      __builtin__.object
 |
 |  Methods defined here:
 |
 |  __init__(self, cmd='clustalw', **kwargs)
 |      # TODO - Should we default to cmd="clustalw2" now?
 |
 |  ----------------------------------------------------------------------
 |  Methods inherited from Bio.Application.AbstractCommandline:
 |
 |  __call__(self, stdin=None, stdout=True, stderr=True, cwd=None,
env=None)
 |      Executes the command, waits for it to finish, and returns output.
 |
 |      Runs the command line tool and waits for it to finish. If it returns
 |      a non-zero error level, an exception is raised. Otherwise two strings
 |      are returned containing stdout and stderr.
 |
 |      The optional stdin argument should be a string of data which will be
 |      passed to the tool as standard input.
 |
```

```
|     The optional stdout and stderr argument may be filenames (string),
|     but otherwise are treated as a booleans, and control if the output
|     should be captured as strings (True, default), or ignored by sending
|     it to /dev/null to avoid wasting memory (False). If sent to a file
|     or ignored, then empty string(s) are returned.
|
|     The optional cwd argument is a string giving the working directory
|         to run the command from. See Python's subprocess module
documentation
|     for more details.
|
|     The optional env argument is a dictionary setting the environment
|     variables to be used in the new process. By default the current
|     process' environment variables are used. See Python's subprocess
|     module documentation for more details.
|
|     Default example usage::
|
|         from Bio.Emboss.Applications import WaterCommandline
|         water_cmd = WaterCommandline(gapopen=10, gapextend=0.5,
|                                      stdout=True, auto=True,
|                                      asequence="a.fasta", bsequence="b.fasta")
|         print("About to run: %s" % water_cmd)
|         std_output, err_output = water_cmd()
|
|     This functionality is similar to subprocess.check_output() added in
|     Python 2.7. In general if you require more control over running the
|     command, use subprocess directly.
|
|     As of Biopython 1.56, when the program called returns a non-zero
error
```

```
|     level, a custom ApplicationError exception is raised. This includes
|     any stdout and stderr strings captured as attributes of the exception
|     object, since they may be useful for diagnosing what went wrong.
|
|  __repr__(self)
|     Return a representation of the command line object for debugging.
|
|     e.g.
|     >>> from Bio.Emboss.Applications import WaterCommandline
|     >>> cline = WaterCommandline(gapopen=10, gapextend=0.5)
|     >>> cline.asequence = "asis:ACCCGGGCGCGGT"
|     >>> cline.bsequence = "asis:ACCCGAGCGCGGT"
|     >>> cline.outfile = "temp_water.txt"
|     >>> print(cline)
|   water -outfile=temp_water.txt -asequence=asis:ACCCGGGCGCGGT
-bsequence=asis:ACCCGAGCGCGGT -gapopen=10 -gapextend=0.5
|     >>> cline
|     WaterCommandline(cmd='water', outfile='temp_
water.txt', asequence='asis:ACCCGGGCGCGGT',
bsequence='asis:ACCCGAGCGCGGT', gapopen=10, gapextend=0.5)
|
|  __setattr__(self, name, value)
|     Set attribute name to value (PRIVATE).
|
|     This code implements a workaround for a user interface issue.
|     Without this __setattr__ attribute-based assignment of parameters
|     will silently accept invalid parameters, leading to known instances
|     of the user assuming that parameters for the application are set,
|     when they are not.
|
|     >>> from Bio.Emboss.Applications import WaterCommandline
```

```
|      >>> cline = WaterCommandline(gapopen=10, gapextend=0.5,
stdout=True)
|    >>> cline.asequence = "a.fasta"
|    >>> cline.bsequence = "b.fasta"
|    >>> cline.csequence = "c.fasta"
|    Traceback (most recent call last):
|    ...
|    ValueError: Option name csequence was not found.
|    >>> print(cline)
|     water -stdout -asequence=a.fasta -bsequence=b.fasta -gapopen=10
-gapextend=0.5
|
|    This workaround uses a whitelist of object attributes, and sets the
|    object attribute list as normal, for these.  Other attributes are
|      assumed to be parameters, and passed to the self.set_parameter
method
|    for validation and assignment.
|
| __str__(self)
|    Make the commandline string with the currently set options.
|
|    e.g.
|    >>> from Bio.Emboss.Applications import WaterCommandline
|    >>> cline = WaterCommandline(gapopen=10, gapextend=0.5)
|    >>> cline.asequence = "asis:ACCCGGGCGCGGT"
|    >>> cline.bsequence = "asis:ACCCGAGCGCGGT"
|    >>> cline.outfile = "temp_water.txt"
|    >>> print(cline)
|  water -outfile=temp_water.txt -asequence=asis:ACCCGGGCGCGGT
-bsequence=asis:ACCCGAGCGCGGT -gapopen=10 -gapextend=0.5
|    >>> str(cline)
```

```
| 'water-outfile=temp_water.txt-asequence=asis:ACCCGGGCGCGGT
-bsequence=asis:ACCCGAGCGCGGT -gapopen=10 -gapextend=0.5'
|
| set_parameter(self, name, value=None)
|     Set a commandline option for a program (OBSOLETE).
|
|     Every parameter is available via a property and as a named
|     keyword when creating the instance. Using either of these is
|     preferred to this legacy set_parameter method which is now
|     OBSOLETE, and likely to be DEPRECATED and later REMOVED
in
|     future releases.
|
| ----------------------------------------------------------------------
| Data descriptors inherited from Bio.Application.AbstractCommandline:
|
| __dict__
|     dictionary for instance variables (if defined)
|
| __weakref__
|     list of weak references to the object (if defined)
|
| ----------------------------------------------------------------------
| Data and other attributes inherited from Bio.Application.
AbstractCommandline:
|
| parameters = None
```

For the most basic usage, all you need is to have a FASTA input file. By default ClustalW will generate an alignment and guide tree file with names based on the input FASTA file, in this case sample.aln and sample.dnd, but you can override this or make it explicit:

```
from Bio.Align.Applications import ClustalwCommandline
cline = ClustalwCommandline("clustalw2", infile="data/sample.fasta")
print(cline)
clustalw2 -infile=data/sample.fasta
```

Notice here we have given the executable name as clustalw2, indicating we have version two installed, which has a different filename to version one (clustalw, the default). Fortunately both versions support the same set of arguments at the command line (and indeed, should be functionally identical). The sample.fasta is any FASTA of your interest, we have left it as text.

You may find that even though you have ClustalW installed, the above command doesn't work – you may get a message about "command not found" (especially on Windows). This indicated that the ClustalW executable is not on your PATH (an environment variable, a list of directories to be searched). You can either update your PATH setting to include the location of your copy of ClustalW tools (how you do this will depend on your OS or operating system), or simply type in the full path of the tool.

The heuristic used in ClustalW is based on phylogenetic analysis. It can be understood to comprise four steps. These steps are described in simple terms below.

- A *pairwise* distance matrix for all the sequences to be aligned is generated. A guide tree is created using the neighbour-joining algorithm.
- Each of the most closely related pairs of sequences - the *outermost branches* of the tree - are aligned to each other using dynamic programming.
- Each new alignment is analyzed to *build* a *sequence profile.*
- Finally, *alignment profiles are aligned* to each other or to other sequences (depending on the topology of the tree) until a full alignment is built.

This strategy produces reasonable alignments under a range of conditions. It's not foolproof; for distantly related sequences, it can build on the inaccuracies of pairwise alignment and phylogenetic analysis. But for sequence sets with some recognizably related pairs, it builds on the strengths of these methods. Pairwise sequence alignment by dynamic programming is very accurate for closely related sequences regardless of which scoring matrix or penalty values are used. Phylogenetic analysis is relatively unambiguous for closely related sequences. Using multiple sequences to create profiles increases the accuracy of pairwise alignment for more distantly related sequences. There are many parameters involved in multiple sequence alignment. There are, of course, scoring matrices and gap penalties associated with the pairwise alignment steps. In addition, there are weighting parameters that alter the scoring matrix used in *sequence-profile* and *profile-profile* alignments. In ClustalW, these are set from the Multiple Alignment submenu or the Profile Structure Alignments submenu. In ClustalX, they are set from the Alignment pulldown menu. The pairwise alignment parameters are familiar and have the same meaning in multiple alignment as they do in pairwise alignment. The multiple alignment parameters include gap opening and gap extension penalties for the multiple alignment process—to be used when fine-tuning alignments—and a maximum allowable delay, in terms of sequence length, for the start of divergent sequences at the beginning of the alignment.

One of ClustalW's heuristics is that, in protein sequence alignment, different scoring matrices are used for each alignment based on expected evolutionary distance. If two sequences are close neighbors in the tree, a scoring matrix optimized for close relationships aligns them. Distant neighbors are aligned using matrices optimized for distant relationships. Thus, when prompted to choose a series of matrices in the Multiple Alignment Parameters menu, it means just that: use BLOSUM62 for close relationships and BLOSUM45 for more distant relationships, rather than the same scoring matrix for all pairwise alignments. Another heuristic that ClustalW uses is *scalable gap penalties* for protein profile alignments. A gap opening next to a conserved hydrophobic residue can be penalized more heavily than a gap opening next to a hydrophilic residue. A gap opening too close to another gap can be penalized more heavily than an isolated gap. These parameters are set in the Protein Gap Parameters menu.

While ClustalW is simple to install and use on a Linux workstation, ClustalX, the X Windows-based graphical user interface for ClustalW, isn't so easy to compile. However, ClustalX runs in its own X window, has pull down menus, and allows viewing and plotting of multiple sequence alignments in a color-coded format. It also allows you to append sequences to an alignment one at a time, and to produce color PostScript output of specified sequence ranges in an alignment from different files, if desired, along with other convenient features. To install ClustalX on a Linux machine, you need:

- The ClustalX binaries
- The NCBI software toolkit source distribution
- The LessTif libraries

SUMMARY

We began our study by looking at the comparison of just two sequences. An algorithm design technique called dynamic programming can be successfully applied in this case, leading to precise and efficient algorithms. We then studied several extensions to the basic dynamic programming algorithms. Some of these extensions allow us to save on the amount of time or space needed for the computation in certain cases. Others are aimed at solving more general forms of the problem necessary in some biological applications. The following is a brief reminder of the various types of comparison and their uses:

- *Global* : compares sequences in its entirety
- *Local*: finds similar pieces in sequences

The multiple-sequence case was studied next. A dynamic programming approach here led to an inefficient algorithm, so other techniques had to be considered; we described a number of these alternative techniques. Another situation in which dynamic programming is not fast enough is when we want to search a database for a given sequence or pattern.

REFERENCES

1. Bioinformatics Programming - Using Python by Mitchell L. Model (O'Reilly, **2010**)
2. Python Programming for Biology by Tim J. Stevens and Wayne Boucher (Cambridge University Press, **2015**).

3. Introduction to Computational Molecular Biology by Joao Setubal and Joao Meidanis (PWS Publishing Company, **1997**)

6

GENOMICS

In the previous chapter you learnt in great detail the methods employed for comparing sequences. We turn our attention in this chapter on Next Generation Sequencing (NGS) and devise some simple recipes for handling the data generated by various genomics and transcriptomics experiments. Again, our focus will be on computational methods.

NEXT GENERATION SEQUENCING

Next-generation Sequencing (NGS) is one of the fundamental technological developments of the decade in life sciences. Whole Genome Sequencing (WGS), RAD-Seq, RNA-Seq, Chip-Seq, and several other technologies are routinely used to investigate important biological problems. These are also called high-throughput sequencing technologies with good reason: they generate vast amounts of data that need to be processed. NGS is the main reason for computational biology becoming a "big data" discipline. More than anything else, this is a field that requires strong bioinformatics techniques. There is a very strong demand for professionals with these skillsets

Here, we will not discuss each individual NGS technique per se (this will require a whole book on its own). We will use an existing WGS dataset and the human 1000 genomes project to illustrate the most common steps necessary to analyze genomic data. The recipes presented here will be easily applicable for other genomic sequencing approaches. Some of them can also be used for transcriptomic analysis (for example, RNA-Seq). The recipes are also species-independent. So, you will be able to apply them to any other species for which you have sequenced data. The biggest difference in processing data from different species is related to genome size, diversity, and the quality of the assembled genome (if it exists for your species). These will not affect the automated Python part of NGS processing much.

In this chapter we also discuss the computational task involved in sequencing, which is called *fragment assembly*. The motivation for this problem comes from the fact that with current technology it is impossible to sequence directly contiguous stretches of more than a few hundred bases. On the other hand, there is technology to cut random pieces of a long DNA molecule and to produce enough copies of the pieces to sequence. Thus, a typical approach to sequencing long DNA molecules is to sample and then sequence fragments from them. However, this leaves us with the problem of assembling the pieces, which is the problem we study in this chapter.

Although you may have your own data to analyze, you will probably need existing genomic datasets. Here, we will see how to access these databases at the National Center for Biotechnology Information (NCBI). We will not only discuss GenBank, but also other databases at NCBI. Many people refer (wrongly) to the whole set of NCBI databases as GenBank, but NCBI includes the nucleotide database and many others, for example, PubMed. As sequencing analysis is a long subject and this book targets intermediate to advanced users, we will not be very exhaustive with a topic that is, at its core, not very complicated. Nonetheless, it's a good warm-up for more complex recipes at the end of this chapter.

FRAGMENT ASSEMBLY

We now present biological background for fragment assembly and then discuss issues related to computational methods used for achieving a meaningful assembly.

To **sequence** a DNA molecule is to obtain the string of bases that it contains. In large scale DNA sequencing we have a long target DNA molecule (typically thousands of bp) that we want to sequence. We may think of this problem as a puzzle in which we are given a double row of cards facing the table, as in Figure 3.1. We do not know which letter from the set {A, C, G, T} is written on each card, but we do know that cards in the same position of opposite strands form a complementary pair. Our goal is

to obtain the letters using certain *hints,* which are (approximate) substrings of the rows.

The long sequence to reconstruct is called the **target.** In the biological problem, we know the length of the target sequence approximately, within 10% or so. It is impossible to sequence the whole molecule directly. However, we may instead get a piece of the molecule starting at a random position in one of the strands and sequence it in the canonical (5' ® 3') direction for a certain length. Each such sequence is called a **fragment.**

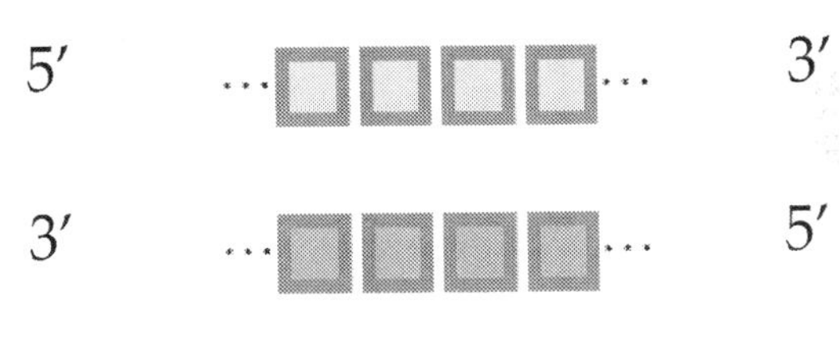

Figure 6.1: Some hypothetical DNA to be sequenced

It corresponds to a substring of one of the strands of the target molecule, *but we do not know which strand or its position relative to the beginning of the strand in addition it may contain errors.*

By using the *shotgun method*, we obtain a large number of fragments and then we try to reconstruct the target molecule's sequence based on fragment overlap. Depending on experimental factors, fragment length can be as low as 200 or as high as 700. Typical problems involve target sequences 30,000 to 100,000 base-pairs long, and total number of fragments is in the range 500 to 2000.

The problem is then to deduce the whole sequence of the target DNA molecule. Because we have a collection of fragments to put together, this task is known as **fragment assembly.** We note that it suffices to determine one of the strands of the original molecule, since the other can be readily obtained because of the complementary pair rule. Typically, the assembly can be thought of starting with the appending of available sequences by *overlaps* between fragments. But, this is compounded by many factors. The real problem instances are very large. Apart from this fact, several other complications exist that make the problem much harder. The main factors that add to the complexity of the problem are *errors*, *unknown orientation*, *repeated regions*, and *lack of coverage*. Let us look at each of these concepts in some detail.

Errors

The simplest errors are called *base call* errors and comprise base *substitutions*, *insertions*, and *deletions* in the fragments. *Transpositions* are also common but we can treat them as compositions of an insertion and a deletion or two substitutions. These are illustrated below, Figures 3.2, 3.3 and 3.4, with some simple examples.

Input:	Answer:
TGGCA	--TGGCA--
GCACG	----GCACG
AATG	AATG-----
ACGGCA	-ACGGCA--
	AATGGCACG

Figure 6.2: A substitution error in the second position of last fragment, where T is replaced by C.

Let us now look at an example of insertion error.

Input:	Answer:
TGGCA	--TGG-CA--
GCACG	----GCACG

AATG	AATG------
ATGGCA	-ATGG-CA--
	AATGG-CACG

Figure 6.3: Insertion error. Base T in the last fragment is correct. In the alignment, you notice a gap inserted after ATGG of the last fragment.

In the above, the consesus will still be correct because of spaces introduced in the multiple alignment and majority voting. Space in the consensus will be discarded when reporting the answer.

Now, let us what a deletion is in fragment assembly.

Input:	Answer:
TGGCA	--TGGCA--
GCACG	----GCACG
AATG	AATG-----
ATGCA	-ATG-CA--
	AATGGCACG

Figure 6.4: Deletion error. Third (or fourth) base in the last fragment has a deletion.

Base call errors occurs in practice at rates varying from 1 to 5 errors every 100 characters. Their distribution along the sequence is not uniform, as they tend to concentrate towards the 3' end of the fragment. As we can see from the examples, it is still possible to reconstruct the correct consensus even in the presence of errors, but the computer program must be prepared to deal with this possibility.

Apart from *erroneous* base calls, two other types of errors can affect assembly. One of them is the artifact of *chimeric* fragments, and the other is *contamination* by host or vector DNA.

Chimeric fragments, or *chimeras,* arise when two regular fragments from distinct parts of the target molecule join end-to-end to form a fragment that is *not* a contiguous part of the target. These misleading fragments must be recognized as such and removed from the fragment set in a preprocessing stage. An example is shown below.

Input:	Answer:
TGGCA	--TGGCA--
GCACG	----GCACG
AATG	AATG-----
ACGGCA	-ACGGCA--
AATACG	AATGGCACG

Figure 6.5: Chimeras. The last fragment in the input is a chimera. The only way to proceed further in assembly is to remove such chimeras.

Sometimes fragments or parts of fragments that do not have anything to do with the target molecule are present in the input set. This is due to *contamination* from host or vector DNA. *Contamination* is a rather common phenomenon in sequencing experiments. Witness to this is the significant quantity of vector DNA that is present in the community databases. Scientists sometimes fail to screen against the vector sequence prior to assembly and submit a contaminated consensus to the database.

Unknown Orientation

Fragments can come from any of the DNA strands and we generally do not know to which strand a particular fragment belongs to. We do know, however, that whatever the strand the sequence read goes from 5' to 3'. Because of the complementarity and opposite orientation of strands, the fact that a fragment is a substring of one strand is equivalent to the fact that its reverse complement is a substring of the other. As a result, we can think of the input fragments as being all approximate substrings of the *consensus* sought either as given or in reverse complement. An example is given below.

Input:	Orientation:	Answer:
TGGCA	→	--TGGCA--
GCACG	→	----GCACG
AATG	←	AATG-----
ACGGCA	←	-ACGGCA--
TGACT	→	--TGACT--

GACT	→	---GACT--
		AATGGCACG

Figure 3.6: Oreintation is unknown in the above assembly. We have no clue about the orientation of each fragment. Each could be either forward or reverse. ® means fragment as is, and ←symbol means its reverse complement.

Repeats

Repeated regions or **repeats** are sequences that appear two or more times in the target molecule. Short repeats, that is, repeats that can be entirely covered by one fragment, do not pose difficulties. The worst problems are caused by longer repeats. Also, the copies of a repeat do not have to be identical to upset assembly. If the level of similarity between two copies of a repeat is high enough, the differences can be mistaken for base call errors. Remember that the assembler must be prepared to deal with errors, so there is usually some degree of tolerance in overlap detection. We illsutrate repeats below.

The kinds of problems that repeats cause are two-fold. First, if a fragment is totally contained in a repeat, we may have several places to put it in the final alignment, as it may fit reasonably well in the several repeat copies. One could argue that it does not matter where we put it, since in any copy of the same repeat the consensus will be approximately the same. But the point is that, when the copies are not exactly equal, we may weaken the consensus by placing a fragment in the wrong copy.

Figure 6.7: Repeated regions in fragment assembly. The same fragment block X_1 is repeated here as X_2.

Second, repeats can be positioned in such a way as to render assembly inherently *ambiguous*; that is, two or more layouts are compatible with the input fragments and approximate target length at equivalent levels of fitness. We illustrate this concept below.

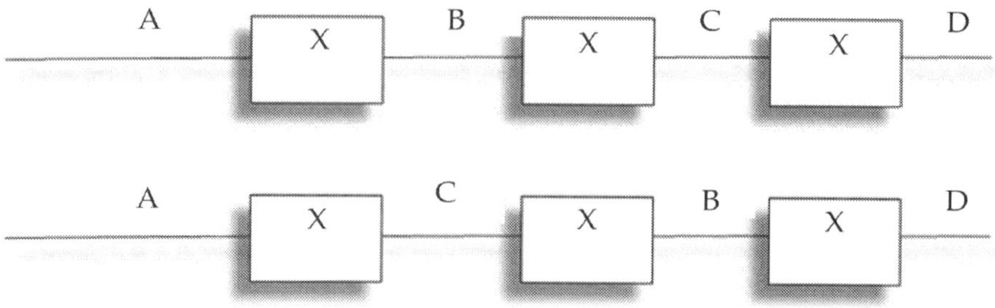

Figure 3.8: Ambiguity in Fragment assembly due to the presence of repeats.

So far we discussed **direct repeats,** namely, repeated copies in the same strand. However, **inverted repeats,** which are repeated regions in opposite strands, can also occur and are potentially more dangerous.

Lack of Coverage

Another problem is lack of adequate coverage. We define the **coverage** at position *i* of the target as the number of fragments that cover this position. This concept is well defined, but it is impossible to compute it because we do not know the actual positions of the fragments in the target. Even after assembly, all we have is our best guess about these positions. We can compute the *mean coverage,* though, by adding up all fragment lengths and dividing by the approximate target length.

If the coverage is equal to zero for one or more positions, then there is not enough information in the fragment set to reconstruct the target completely. Figure 3.2 shows an example in which there is an uncovered area. In such cases, the best we can hope to achieve is a layout for every one of the contiguously covered regions, called **contigs.** There are two contigs in Figure 3.9.

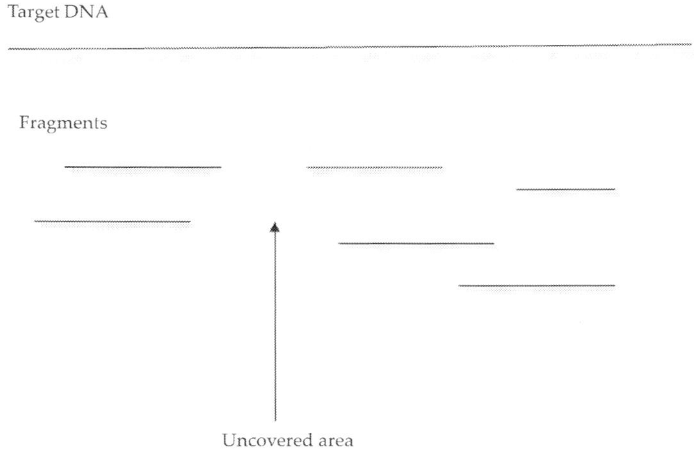

Figure 3.9: Incomplete or Insufficient coverage in a fragment assembly.

Lack of coverage occurs because the sampling of fragments is essentially a random process; it is therefore possible that some parts end up not being well covered or not covered at all. In general, we want not only one, but several different fragments covering any given point. The more fragments we have, the safer is our assessment of the consensus based on voting. It is also desirable to have fragments from both strands covering a given region, for it has been observed in practice that certain kinds of errors occur consistently in one strand only.

Insufficient coverage can usually be dealt with by sampling more fragments, but we must be careful with this approach. If everything is well covered except for a small portion of the target, sampling at random can be a very inefficient way of covering the gaps. An approach, called *directed sequencing* or *walking*. To avoid the formation of these little gaps, some researchers advocate sampling at very high coverage rates, collecting fragments whose combined length is enough to cover the target molecule 8 times or more.

It is important to know how many fragments we need to generate in order to achieve a given coverage. The answer cannot be absolute, of course, since sampling is a random process, but reasonable bounds can be derived. We mention here an important result that is valid under certain simplifying assumptions.

Let T denote the length of the target molecule. Assume that all fragments have about the same length l and that we can safely recognize overlaps of at least t bases. If we sample n fragments at random, the expected number p of *apparent contigs* is given by the formula:

$$p = ne^{-\frac{n(l-t)}{T}}$$

The term "apparent contig" refers to the fact that we are assuming that we are able to recognize overlaps only if they have size at least t. Thus some of the contigs we have may in fact overlap by less that t bases, so they are really parts of a longer true contig but "appear" to be two separated contigs to us.

Notice that p approaches zero as the number n of fragments grows, which seems strange because we know there will always be at least one apparent contig. This happens because formula above is an approximation. It can be used also in the context of DNA mapping, where the number of contigs is much larger, so a difference of one or two is negligible. The formula is useful in the sense that it provides a reliable ballpark in a wide variety of situations.

A formula for the fraction of the target molecule covered by the fragments is also available. With the same notation as above, the fraction covered by exactly k fragments is given by:

$$r_k = e^{-c}\, c^k / k!$$

where

$$c = \frac{nl}{T}$$

is the *mean coverage.*

ALTERNATIVE TO DNA SEQUENCING

We start with *directed sequencing,* a method that can be used to cover small remaining gaps in a shotgun project. In direct sequencing a special primer is derived from the sequence near the end of a contig, so that fragments spanning the region including the end of this contig and the continuation of it in the target are generated. These new fragments are then sequenced and give the sequence adjacent to the contig, thereby augmenting it. Continuing in this fashion, we can cover the gap to the next contig. The problem with this approach is that it is expensive to build special primers. Also, the next step can be accomplished only after the current one, so the process is essentially sequential rather than parallel (but can be done for all gaps in parallel).

Another technique that has become very popular is called *dual end sequencing*. In a shotgun experiment several copies of the target DNA molecule are broken randomly and short pieces are selected for cloning and sequencing. These pieces are called *inserts* because they are inserted into a vector for amplification.

Inserts sizes range from 1 to 5 kbp, but only about 200 to 700 bases can be directly read from one extremity to yield a fragment. However, it is also possible to read the other extremity if we have a suitable primer. The two fragments thus obtained belong to opposite strands, and they should be separated in the final alignment by roughly the insert size minus the fragment size. This extra information is extremely useful in closing gaps, for instance. Sometimes the dual end is sequenced only if it is necessary to close a gap. Notice that dual end sequencing exploits the fact that inserts are usually larger than the portion read from them.

A radically different approach from shotgun methods has been jproposed recently. It is known as *sequencing by hybridization* or **SBH**. As this book is more focussed on computational methods we will not digress into the vast sea of sequencing methodologies which have come up since the human genome project.

Let us start by working with modern sequence formats, as are popular in the field of genomics. Here, we will work with FASTQ files, the standard format output by modern sequencers. You will learn how to work with quality scores per base and also consider the variations in output coming from different sequencing machines and databases. This is the first recipe that will use real data (*big data*) from the human 1000 genomes project. We will start with a brief description of the project.

1000 Genomes Project

The human 1000 genomes project aims to catalog world-wide human genetic variation and takes advantage of modern sequencing technology to do WGS. This project makes all data publicly available, which includes output from sequencers, sequence alignments, and SNP calls, among many other artifacts. The name 1000 genomes is actually a misnomer because it currently includes more than 2500 samples. These samples are divided into 26 populations, spanning the whole planet. We will mostly use data from four populations: African Yorubans (YRI), Utah Residents with Northern and Western European Ancestry (CEU), Japanese in Tokyo (JPT), and Han Chinese in Beijing (CHB). The reason we chose these specific populations is because they were the first ones that came from HapMap, an old project with similar goals. They used genotyping arrays to know more about the quality of this subset.

Download the following file for the hands-on exposure to processing FASTQ files:

ftp://ftp.1000genomes.ebi.ac.uk/vol1/ftp/phase3/data/HG03785/sequence_read/ERR181394.filt.fastq.gz

This is a fairly small file (~ 5.5 MB) and represents part of a sequenced data of an Indian male (HG03785). Before we start coding, let's take a look at the FASTQ file, in which you will have many records, as shown in the following code:

```
@SRR003258.1 30443AAXX:1:1:1053:1999 length=51
ACCCCCCCCCACCCCCCCCCCCCCCCCCCCCCCCCCACACA-
CACCAACAC

+

=IIIIIIII5IIIIIII>IIII+GIIIIIIIIIIIIII(IIIII01&III
```

Line 1 starts with an @, followed by a sequence identifier and a description string. The description string will vary from a sequencer or a database source, but will normally be amenable to automated parsing. The second line has the sequence DNA, which is just like a FASTA file. The third line is a +, sometimes followed by the description line on the first line. The fourth line contains quality values for each base read on line two. Each letter encodes a Phred quality score. Check the wikipedia link below for more information:

http://en.wikipedia.org/wiki/Phred_quality_score

This assigns a *probability of error* to each read. This encoding can vary a bit among platforms. Be sure to check for this on your specific platform. Let's start by reading this file using BioPython.

```
from collections import defaultdict
%matplotlib inline
import seaborn as sns
import matplotlib.pyplot as plt
from Bio import SeqIO
recs = SeqIO.parse('ERR181394.filt.fastq', 'fastq')
rec = next(recs)
print(rec.id, rec.description, rec.seq)
print(rec.letter_annotations)
```

ERR181394.988 ERR181394.988 HS16_08506:4:1101:1445:20696 #61/2 TCGGAGCAGCAGTTTTAGGTTCATAGCCAAACTGAGA GG-GAGATACAGAGATTTCCTATCAGCCCTATAGCCCACAC ATCTGTAACCTCTCCCACTTTCA {'phred_quality': [24, 33, 34, 28, 35, 35, 37, 35, 36, 37, 36, 36, 34, 38, 38, 37, 38, 40, 38, 24, 33, 37, 37, 34, 39, 29, 38, 34, 36, 35, 33, 37, 12, 36, 35, 37, 38, 39, 34, 37, 22, 37, 23, 36, 34, 38, 36, 37, 38, 37, 34, 28, 34, 38, 41, 37, 38, 35, 33, 39, 37, 38, 36, 21, 35, 31, 20, 21, 38, 38, 17, 37, 38, 38, 36, 38, 37, 31, 32, 36, 38, 29, 25, 32, 26, 34, 37, 17, 34, 24, 37, 37, 37, 35, 35, 23, 31, 29, 29, 33]}

```
# Let us count the alphabet in the FASTQ file
recs = SeqIO.parse('ERR181394.filt.fastq', 'fastq')
cnt = defaultdict(int)
for rec in recs:
        for letter in rec.seq:
                cnt[letter] += 1
tot = sum(cnt.values())
for letter, cnt in cnt.items():
print('%s: %.2f %d' % (letter, 100. * cnt / tot, cnt))
```

T: 28.35 1592029

C: 19.99 1122341

G: 20.08 1127472

A: 29.29 1644855

N: 2.30 129203

N's are basically calls in which a sequencer cannot distinguish between the 4 types and flags it as unknown. You must notice that it is a filtered file, hence the proportion of N's is much low !

```
# We try to visualize the N's according to its position in the reads.
recs = SeqIO.parse('ERR181394.filt.fastq', 'fastq')
n_cnt = defaultdict(int)
for rec in recs:
        for i, letter in enumerate(rec.seq):
        pos =i +1
        if letter == 'N':
                n_cnt[pos] += 1
seq_len = max(n_cnt.keys())
 positions = range(1, seq_len + 1)
fig, ax = plt.subplots()
ax.plot(positions, [n_cnt[x] for x in positions])
 ax.set_xlim(1, seq_len)
```

(1, 100)

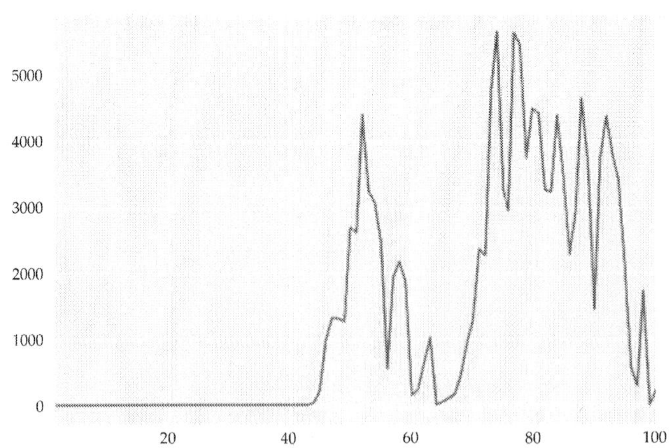

You will notice that there are no N's till 45 or so and after 45th position the distribution is far from uniform. This brings us to the question of:

How about the quality of the reads ? Let's check that now.

```
recs = SeqIO.parse('ERR181394.filt.fastq', 'fastq')
cnt_qual = defaultdict(int)
for rec in recs:
    for i, qual in enumerate(rec.letter_annotations['phred_quality']):
        if i < 45:
            continue
        cnt_qual[qual] += 1
tot = sum(cnt_qual.values())
for qual, cnt in cnt_qual.items():
    print('%d: %.2f %d' % (qual, 100. * cnt / tot, cnt))
38: 15.75 486400
36: 9.60 296505
37: 14.70 454019
34: 6.14 189789
28: 1.85 57284
41: 0.50 15571
35: 6.34 195808
33: 3.81 117807
39: 4.62 142568
21: 1.34 41394
31: 2.03 62779
20: 0.80 24626
17: 1.34 41427
32: 3.14 96872
29: 1.93 59459
25: 0.99 30517
26: 1.52 46851
```

24: 1.05 32377

23: 0.80 24596

19: 2.04 62862

22: 0.59 18071

11: 0.85 26101

6: 0.71 22031

15: 0.51 15862

30: 2.40 73994

10: 2.36 72886

9: 1.39 42812

5: 0.78 24023

16: 0.94 29074

12: 0.39 12163

0: 4.18 128994

40: 0.77 23922

18: 0.96 29628

14: 0.50 15398

8: 0.62 19128

7: 0.31 9691

3: 0.08 2612

27: 1.14 35062

42: 0.03 1009

4: 0.10 3031

43: 0.01 201

1: 0.00 59

13: 0.08 2414

2: 0.03 1068

HANDLING ALIGNMENT DATA

With the data from sequencer in hand, one resorts to tools like bwa to align sequences to a reference genome. The program implements Burrows-Wheeler transform (BWT). It is also called block-sorting compression. Basically, it is a rearrangement of character strings which is done in runs of similar characters. When a character string is transformed by the BWT, the transformation permutes the order of characters. You may wonder how can such an algorithm find a place in NGS work.

We illustrate this with a simple example.

```
def rotations(t):
    # Return list of rotations of input string t
    tt = t * 2
    return [ tt[i:i+len(t)] for i in range(0, len(t)) ]
rotations('cat')
['cat', 'atc', 'tca']
def bwm(t):
    # Return lexicographically sorted list of t's rotations
    return sorted(rotations(t))
bwm('abaaba$')
['$abaaba', 'a$abaab', 'aaba$ab', 'aba$aba', 'abaaba$', 'ba$abaa', 'baaba$a']
print('\n'.join(bwm('abaaba$')))
$abaaba
a$abaab
aaba$ab
aba$aba
abaaba$
ba$abaa
baaba$a
def bwtViaBwm(t):
```

```
    # Given T, returns BWT(T) by way of the BWM
    return ''.join(map(lambda x: x[-1], bwm(t)))
bwtViaBwm('abaaba$') # we can see the result equals the last column of the matrix above
'abba$aa'
def suffixArray(s):
    satups = sorted([(s[i:], i) for i in range(0, len(s))])
    return map(lambda x: x[1], satups)
def bwtViaSa(t):
    # Given T, returns BWT(T) by way of the suffix array
    bw = []
    for si in suffixArray(t):
        if si == 0:
            bw.append('$')
        else:
            bw.append(t[si-1])
    return ''.join(bw) # return string-ized version of list bw
bwtViaBwm('abaaba$'), bwtViaSa('abaaba$') # same result
('abba$aa', 'abba$aa')
```

In NGS, DNA is fragmented into small pieces, of which the first few bases are sequenced, yielding several millions of "reads", each 30 to 500 base pairs ("DNA characters") long. In many experiments, e.g., in ChIP-Seq, the task is now to align these reads to a reference genome, i.e., to the known, nearly complete sequence of the organism in question (which may be up to several billion base pairs long). A number of alignment programs, specialized for this task, were published, which initially relied on hashing. In an effort to reduce the memory requirement for sequence alignment, several alignment programs were developed (like bwa). We will not go into the particulars, rather we jump into handling the alignment data itself.

The aligned data is represented as Sequence Alignment/Map (SAM) format. Due to the massive size of most of these files, you will probably work with its compressed version (BAM). The compressed format is indexable for extremely fast random access (for example, to speedily find alignments to a certain part of a chromosome). Note that you will need to have an index for your BAM file normally created by the tabix utility of samtools. Samtools is probably the most widely used tool to manipulate SAM/BAM files.

We will use the chromosome data from the previous example. This is a paired end data, with reads of 76 bp. Adjust according to your data. We will be using Pysam which is a wrapper to the samtools C API. You know how to install these tools by now. You can follow the tips given in the earlier chapters and take a peek at the header of the data. We do that below.

```
import pysam
bam = pysam.AlignmentFile('HG03785.chrom20.ILLUMINA.bwa.
ITU.exome.20130415.bam', 'rb')
headers = bam.header
for record_type, records in headers.items():
        print (record_type)
        for i, record in enumerate(records):
                print ('\t%d' % (i + 1))
                #print (record)
if type(record) is dict:
        for field, value in record.items():
                print('\t\t%s\t%s' % (field, value))
```

The header is represented as a dictionary. As there can be several instances of the same record type, the value of the dictionary is a list.

Phew! That was quite a lot of output. I have shown only the first few lines above. I am sure, the reader would have skipped the pages to see if this makes any sense at all. Let us inspect a single record to understand the above information. For this we need to do the following.

```
#0-based
for rec in bam:
        if rec.cigarstring.find('M') > -1 and rec.cigarstring.find('S') > -1
```

and

not rec.is_unmapped and not rec.mate_is_unmapped:

break

print(rec.query_name, rec.reference_id, bam.getrname(rec.reference_id),

rec.reference_start, rec.reference_end)

print(rec.cigarstring)

print(rec.query_alignment_start, rec.query_alignment_end, rec.query_alignment_length)

print(rec.next_reference_id, rec.next_reference_start, rec.template_length)

print(rec.is_paired, rec.is_proper_pair, rec.is_unmapped, rec.mapping_quality)

print(rec.query_qualities)

print(rec.query_alignment_qualities)

print(rec.query_sequence)

Let us see how the record looks like:

ERR250759.22349833 19 20 59992 60067

25S75M

25 100 75

19 60187 285

True True False 29

array('B', [25, 23, 31, 30, 31, 27, 31, 32, 33, 37, 30, 32, 33, 38, 29, 29, 25, 31, 28, 29, 38, 30, 38, 29, 36, 37, 30, 36, 31, 36, 34, 35, 30, 37, 30, 29, 28, 33, 31, 39, 26, 37, 34, 4, 31, 30, 31, 30, 36, 32, 36, 36, 31, 31, 39, 27, 24, 31, 26, 23, 31, 26, 30, 22, 26, 31, 29, 33, 33, 34, 29, 29, 29, 13, 29, 28, 29, 25, 29, 29, 34, 37, 31, 30, 32, 30, 35, 33, 28, 26, 31, 31, 34, 33, 30, 34, 30, 28, 28, 33])

array('B', [37, 30, 36, 31, 36, 34, 35, 30, 37, 30, 29, 28, 33, 31, 39, 26, 37, 34, 4, 31, 30, 31, 30, 36, 32, 36, 36, 31, 31, 39, 27, 24, 31, 26, 23, 31, 26, 30, 22, 26, 31, 29, 33, 33, 34, 29, 29, 29, 13, 29, 28, 29, 25, 29,

29, 34, 37, 31, 30, 32, 30, 35, 33, 28, 26, 31, 31, 34, 33, 30, 34, 30, 28, 28, 33])

GAAAAAACTGAAAGTTAATAGAGAGGTGACTC-AGATCCA GAGGGGGAAGAGGAAGGAAGCTTGGAACCCTA-TAGAGT TGCTGAGTGCCAGGACCAGATCC

- Note that the BAM file object is iterable over its records. We will transverse it until we find a record whose CIGAR string contains a match and a soft clip.
- The CIGAR string gives an indication of the alignment of individual bases. The clipped part of the sequence is a part that the aligner failed to align (but is not removed from the sequence). We will also want the read, its mate ID, and position (the pair, as we have paired-end reads) that was mapped to the reference genome.
- First, we print the query template name followed by the reference ID. The reference ID is a pointer to name of the sequence on the given references on the lookup table of references. An example will make this clear. For all records on this BAM file, the reference ID is 19 (a noninformative number), but if you apply bam.getrname(19), you will get 20, which is the name of the chromosome. So, do not confuse the reference ID (in this case, 19) with the name of the chromosome (20). This is then followed by the reference start and reference end. *Pysam* is 0-based, not 1-based. So, be careful when you convert coordinates to other libraries.
- Then, we print where the alignment starts and ends and calculate its length. By the way, you can compute this by looking at the CIGAR string. It starts at 0 (as the first part of the read is mapped) and ends at 52. The length is 76 again.
- Now, we query the mate (something that you will only do if you have paired-end reads). We get its reference ID (as shown in the previous code), its start position, and a measure of the distance between both pairs. This measure of distance only makes sense if both mates are mapped to the same chromosome.

- We then plot the Phred score (refer to the previous recipe on Phred scores) for the sequence, and then only for the aligned part.
- Finally, we print the sequence (not to forget this!). This is the complete sequence, not the clipped one (of course, you can use the preceding coordinates to clip).

For more information on formats the reader is referred to the following:

- The SAM/BAM format is described at http://samtools.github.io
- If you really need to get complex statistics from BAM files, Alistair Miles' pysamstats
- library is your port of call at https://github.com/alimanfoo/pysamstats
- The technical forum for sequence analysis is SEQanswers (http://seqanswers.com/)

We provide a desscription of SAM format and FASTQ in the Appendix 1.

SUMMARY

We have introduced the reader to some of the common methods and tools to handle genomics data. We introduce advanced topics in the next chapter.

REFERENCES

1. Introduction to Computational Genomics: A Case Studies Approach by Cristianini N & Hahn MW (Cambridge University Press, 2006)
2. Molecular Evolution: A Phylogenetic Approach by Page RDM & Holmes EC (Wiley-Blackwell, 2009).
3. An Introduction to Applied Bioinformatics by Greg Caporaso.

4. Altschul S, Warren G, Miller Webb, Myers E, Lipman D. "Basic Local Alignment Search Tool" J. Mol. Biol. 215 (3): 403-410 (1990).
5. Lipman DJ, Pearson WR. "Rapid and Sensitive Protein Similarity Searches" Science 227 (4693): 1435-1441 (1985)

7

GENOMICS DATA ANALYSIS

Many tasks in computational biology are dependent on the existence of **reference genomes**. If you are performing a sequence alignment, finding genes, or studying genetics of populations at several points of your work, you will be directly or indirectly using a genome reference. In this chapter, we will develop some recipes to work with reference genomes and deal with varying quality of references (which can vary for high-quality, like with the human genome, to problematic with non-model species). We will also see how to deal with genome annotations (working with text databases that will point us to interesting features in the genome) and extract sequence data using the annotation information. Also, we will try to find some gene orthologues across species.

HIGH QUALITY GENOME

In this recipe, you will learn a few general techniques to manipulate reference genomes. As an illustrative example, we will study the GC content (the fraction of the genome that is based on Guanine-Cytosine). Reference genomes are normally made available as FASTA files.

Genomes come in widely different sizes, ranging from viruses such as HIV (which is 9.7 kbp) to bacteria such as *E. coli*, to protozoans such as *Plasmodium falciparum* (the most important parasite species causing malaria) with its 14 chromosomes, mitochondrion, and apicoplast, to the fruit fly with three autosomes, a mitochondrion, and X/Y sex chromosomes, to humans with its three Gbp pairs spread across 22 autosomes, X/Y chromosomes, and mitochondria, all the way up to *Paris japonica*, a plant with 150 Gbp of genome. Along the way, you have different ploidy and different sex chromosome organizations.

In order to make this recipe less burdensome, we will use a small eukaryotic genome from *Plasmodium falciparum*. This genome still has many typical features of larger genomes (for example, multiple chromosomes). So, it's a good compromise between complexity and size. Note that with a genome of the size of *P. falciparum*, it will be possible to perform many operations by loading the whole genome in-memory. However, we opted for a programming style that can be used with bigger genomes (for example, mammals) so that you can use this recipe in a more general way, but feel free to use more memory-intensive approaches with small genomes like this. We will use Biopython.

Let us get started with exploring a reference genome now.

We start by inspecting the description of all the sequences on the reference genome FASTA file:

```
From Bio import SeqIO
genome_name = ‘PlasmoDB-9.3_Pfalciparum3D7_Genome.fasta’
recs = SeqIO.parse(genome_name, ‘fasta’)
chroms = {}
for rec in recs:
        print(rec.description)
Pf3D7_05_v3 | organism=Plasmodium_falciparum_3D7 | version=2012-02-01 |
length=1343557 | SO=chromosome
Pf3D7_10_v3 | organism=Plasmodium_falciparum_3D7 | version=2012-02-01 |
length=1687656 | SO=chromosome
```

Pf3D7_07_v3 | organism=Plasmodium_falciparum_3D7 | version=2012-02-01 |

length=1445207 | SO=chromosome

Pf3D7_03_v3 | organism=Plasmodium_falciparum_3D7 | version=2012-02-01 |

length=1067971 | SO=chromosome

Pf3D7_13_v3 | organism=Plasmodium_falciparum_3D7 | version=2012-02-01 |

length=2925236 | SO=chromosome

Pf3D7_11_v3 | organism=Plasmodium_falciparum_3D7 | version=2012-02-01 |

length=2038340 | SO=chromosome

Pf3D7_14_v3 | organism=Plasmodium_falciparum_3D7 | version=2012-02-01 |

length=3291936 | SO=chromosome

Pf3D7_09_v3 | organism=Plasmodium_falciparum_3D7 | version=2012-02-01 |

length=1541735 | SO=chromosome

Pf3D7_01_v3 | organism=Plasmodium_falciparum_3D7 | version=2012-02-01 |

length=640851 | SO=chromosome

Pf3D7_12_v3 | organism=Plasmodium_falciparum_3D7 | version=2012-02-01 |

length=2271494 | SO=chromosome

Pf3D7_08_v3 | organism=Plasmodium_falciparum_3D7 | version=2012-02-01 |

length=1472805 | SO=chromosome

Pf3D7_06_v3 | organism=Plasmodium_falciparum_3D7 | version=2012-02-01 |

length=1418242 | SO=chromosome

Pf3D7_04_v3 | organism=Plasmodium_falciparum_3D7 | version=2012-02-01 |

length=1200490 | SO=chromosome

Pf3D7_02_v3 | organism=Plasmodium_falciparum_3D7 | version=2012-02-01 |

length=947102 | SO=chromosome

M76611 | organism=Plasmodium_falciparum_3D7 | version=2012-02-01 |

length=5967 | SO=mitochondrial_chromosome

PFC10_API_IRAB | organism=Plasmodium_falciparum_3D7 | version=2012-02-01 |

length=34242 | SO=apicoplast_chromosome

In this example, you can see that we have chromosomes, mitochondria, and apicoplast. We can also view chromosome sizes, but we will take the value from the sequence length instead.

Let's parse the description line to extract the chromosome number. We will retrieve the chromosome size from the sequence and compute the GC content across chromosomes on a window basis:

```
from __future__ import print_function
from Bio import SeqUtils
recs = SeqIO.parse(genome_name, 'fasta')
chrom_sizes = {}
chrom_GC = {}
block_size = 50000
min_GC = 100.0
max_GC = 0.0
for rec in recs:
    if rec.description.find('SO=chromosome') == -1:
        continue
chrom = int(rec.description.split('_')[1])
chrom_GC[chrom] = []
size = len(rec.seq)
chrom_sizes[chrom] = size
num_blocks = size // block_size + 1
```

```
for block in range(num_blocks):
    start = block_size * block
    if block == num_blocks - 1:
        end = size
    else:
        end = block_size + start + 1
    block_seq = rec.seq[start:end]
    block_GC = SeqUtils.GC(block_seq)
    if block_GC < min_GC:
        min_GC = block_GC
    if block_GC > max_GC:
        max_GC = block_GC
    chrom_GC[chrom].append(block_GC)
print(min_GC, max_GC)
16.14167716645667 43.673469387755105
```

We start by defining a window size of 50 kbp. This is appropriate for *P. falciparum* (feel free to vary its size), but you will want to consider other values for genomes with chromosomes that are orders of magnitude different from this.

Note that we are re-reading the file. With such a small genome, it would have been feasible (in step one) to do an in-memory load of the whole genome. By all means, feel free to try this programming style for small genomes - it's faster! However, this code is more generalized for larger genomes.

Note that in the for loop, we ignore the mitochondrion and apicoplast by parsing the SO entry to the description. The chrom_sizes dictionary will maintain the size of chromosomes.

The chrom_GC dictionary is our most interesting data structure and will have a list of faction of the GC content for each 50 kbp window. So, for chromosome 1, which has a size of 640,851 bp, there will be 14 entries because this chromosome size has 14 blocks of 50 kbp.

Now, let's perform a genome plot of the GC distribution. We will use shades of blue for the GC content. However, for high outliers, we will use shades of red. For low outliers, we will use shades of yellow:

```
from __future__ import division
from reportlab.lib import colors
from reportlab.lib.units import cm
from Bio.Graphics import BasicChromosome
```

We will use float division and import functions required by Biopython from the *reportlab* library. Reader is directed to install it in their computing environment.

```
chroms = list(chrom_sizes.keys())
chroms.sort()
biggest_chrom = max(chrom_sizes.values())
my_genome = BasicChromosome.Organism(output_format='png')
my_genome.page_size = (29.7*cm, 21*cm)
telomere_length = 10
bottom_GC = 17.5
top_GC = 22.0
for chrom in chroms:
    chrom_size = chrom_sizes[chrom]
    chrom_representation = BasicChromosome.Chromosome \
        ('Cr %d' % chrom)
    chrom_representation.scale_num = biggest_chrom
    tel = BasicChromosome.TelomereSegment()
    tel.scale = telomere_length
    chrom_representation.add(tel)
    num_blocks = len(chrom_GC[chrom])
    for block, gc in enumerate(chrom_GC[chrom]):
        my_GC = chrom_GC[chrom][block]
```

```
            body = BasicChromosome.ChromosomeSegment()
            if my_GC > top_GC:
                    body.fill_color = colors.Color(1, 0, 0)
            elif my_GC < bottom_GC:
                    body.fill_color = colors.Color(1, 1, 0)
            else:
                    my_color = (my_GC - bottom_GC) / (top_GC -
bottom_GC)
                    body.fill_color = colors.Color(my_color, my_
color, 1)
            if block < num_blocks - 1:
                    body.scale = block_size
            else:
                    body.scale = chrom_size % block_size
            chrom_representation.add(body)
        tel = BasicChromosome.TelomereSegment(inverted=True)
        tel.scale = telomere_length
        chrom_representation.add(tel)
        my_genome.add(chrom_representation)
my_genome.draw('falciparum.png', 'Plasmodium falciparum')
```

The first line converts the return of the keys method to a list. This is redundant in Python 2, but not in Python 3, where the keys method has a specific return type: dict_keys.

We will draw the chromosomes in order (hence the sort). We will need the size of the biggest chromosome (14 in *P. falciparum*) in order to assure that the size of chromosomes is printed with the correct scale (the biggest_chrom variable).

We then create an A4-sized representation of an organism with a PNG output. Note that we will draw very small telomeres of 10 bp. This will produce a rectangular-like chromosome. You can make the telomeres bigger, giving it a roundish representation (or you may have a better idea of the correct telomere size for your species).

We declare that anything with GC content below 17.5 percent or above 22.0 percent will be considered an outlier. Remember that for most other species, this will be much higher.

We then print these chromosomes proper. They are bounded by telomeres and composed of 50 kbp chromosome segments (the last segment is sized with the remainder). Each segment will be colored in blue with a red-green component based on the linear normalization between two outlier values. Each chromosome segment will either be 50 kbp or potentially smaller if the last one of the chromosome. The output is shown in the following figure:

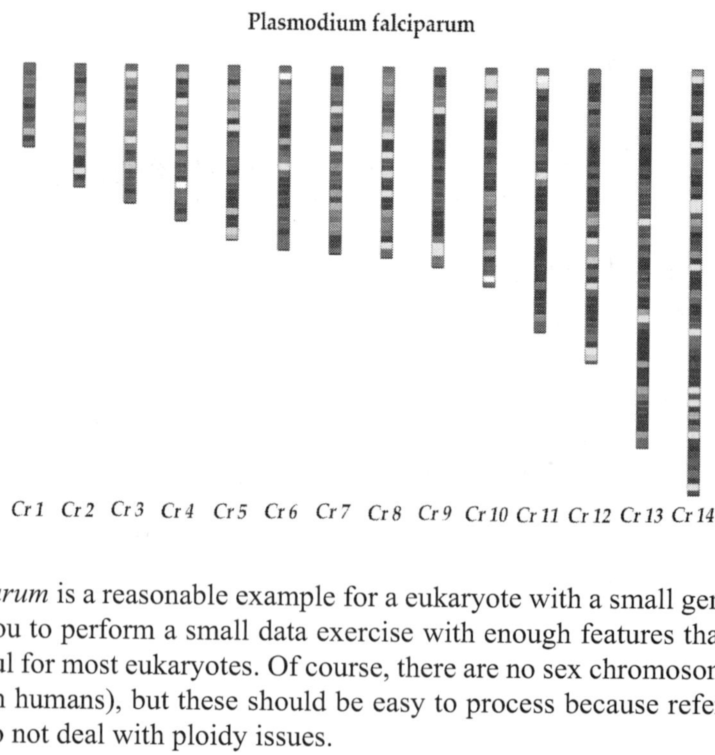

P. falciparum is a reasonable example for a eukaryote with a small genome that allows you to perform a small data exercise with enough features that make it still useful for most eukaryotes. Of course, there are no sex chromosomes (such as X/Y in humans), but these should be easy to process because reference genomes do not deal with ploidy issues.

Biopython does have the functionality to print circular genomes, which you can also use with bacteria. With regards to bacteria and viruses, these genomes are much easier to process because their size is very small.

You can find many reference genomes of model organisms in Ensembl at

http://www.ensembl.org/info/data/ftp/index.html

NCBI also provides a large list of genomes at

http://www.ncbi.nlm.nih.gov/genome/browse/

LOW QUALITY GENOME

Unfortunately, not all reference genomes will have the quality of *P. falciparum*. In keeping with the malaria theme, here, we will use the reference genomes of two mosquitoes that are vectors of malaria:

- *Anopheles gambiae* (which is the most important vector of malaria and can be found in sub-Saharan Africa) and
- *Anopheles atroparvus*, a malaria vector in Europe (while the disease has been eradicated in Europe, this vector is still around).

The *An. gambiae* genome is of reasonable quality. Most chromosomes have been mapped, although the Y chromosome still needs some work. There is a fairly large “unknown” chromosome, probably composed of bits X and Y chromosomes and also midgut microbiota. This genome has a reasonable amount of positions that are not called (that is, you will find Ns instead of ACTGs). The *An. atroparvus* genome is still in the scaffold format. Unfortunately, this is what you will find for many non-model species.

The Anopheles genome is one order of magnitude bigger than the *P. falciparum* genome (but one order of magnitude smaller than most mammals).

We begin by listing the chromosomes in A. gambiae genome:

```
from __future__ import print_function
import gzip
from Bio import SeqIO, SeqUtils
gambiae_name ='gambiae.fa'
atroparvus_name ='atroparvus.fa'
recs = SeqIO.parse(gambiae_name, 'fasta')
for rec in recs:
    print(rec.description)
chromosome:AgamP3:2L:1:49364325:1 chromosome 2L
chromosome:AgamP3:2R:1:61545105:1 chromosome 2R
chromosome:AgamP3:3L:1:41963435:1 chromosome 3L
chromosome:AgamP3:3R:1:53200684:1 chromosome 3R
chromosome:AgamP3:UNKN:1:42389979:1 chromosome UNKN
chromosome:AgamP3:X:1:24393108:1 chromosome X
chromosome:AgamP3:Y_unplaced:1:237045:1 chromosome Y_unplaced
```

The code is quite straightforward. We can see four chromosome arms (2L, 2R, 3L, and 3R) and the X chromosome. Note the Y chromosome, which is quite small and has a name that all but indicates that it might not be in the best state. Also, note that the unknown (UNKN) chromosome is actually a quite large proportion of the reference genome to the tune of the chromosome arm.

We will now check uncalled positions (Ns) and its distribution for the An. gambiae genome:

```
from __future__ import division
recs = SeqIO.parse(gambiae_name, 'fasta')
chrom_Ns = {}
chrom_sizes = {}
for rec in recs:
```

```
    chrom = rec.description.split(':')[2]
    if chrom in ['UNKN', 'Y_unplaced']:
        continue
    chrom_Ns[chrom] = []
    on_N =False
    curr_size =0
for pos, nuc in enumerate(rec.seq):
        if nuc in ['N', 'n']:
                        curr_size +=1
                on_N =True
        else:
                if on_N:
                        chrom_Ns[chrom].append(curr_size)
                        curr_size =0
                                on_N =False
        if on_N:
                chrom_Ns[chrom].append(curr_size)
                chrom_sizes[chrom] = len(rec.seq)
for chrom, Ns in chrom_Ns.items():
    size = chrom_sizes[chrom]
    print('%s (%s): %%Ns (%.1f), num Ns: %d, max N: %d'% (chrom,
size, 100* sum(Ns) / size, len(Ns), max(Ns)))
2L (49364325): %Ns (1.7), num Ns: 957, max N: 28884
2R (61545105): %Ns (2.3), num Ns: 1658, max N: 36427
3L (41963435): %Ns (2.9), num Ns: 1272, max N: 31063
3R (53200684): %Ns (1.8), num Ns: 1128, max N: 24292
X (24393108): %Ns (4.1), num Ns: 1287, max N: 21132
```

This code will take some time to run, so please be patient. We will inspect each and every base pair of the autosomes. Hence, re-open and re-read the file to save memory.

We have two dictionaries: one dictionary with chromosome sizes and another with the distribution of the sizes of runs of Ns. To calculate the runs of Ns, we traverse all autosomes (noting when an N position starts and ends). We then print the basic statistics of the distribution of Ns.

So, for the 2L chromosome arm (with a size of 49 Mbp), 1.7 percent are N calls divided by 957 runs. The biggest run is 28,884 bps. Note that the X chromosome has the highest fraction of positions with Ns.

Let us look at *An. Atroparvus* genome now. We count the number of saffolds along with the distriution of scaffold sizes.

```
import numpy as np
recs = SeqIO.parse(atroparvus_name, 'fasta')
sizes = []
size_N = []
for rec in recs:
    size = len(rec.seq)
    sizes.append(size)
    count_N =0
for nuc in rec.seq:
        if nuc in ['n', 'N']:
                count_N +=1
        size_N.append((size, count_N / size))
print(len(sizes), np.median(sizes), np.mean(sizes), max(sizes), min(sizes),
    np.percentile(sizes, 10), np.percentile(sizes, 90))
1371 8304.0 163596.006565 20238125 1004 1563.0 56612.0
```

We thus have 1,371 scaffolds (against seven entries on the A. gambiae genome) with a median size of 8,304 (mean of 163,536). The biggest scaffold has 2 Mbp and the smallest scaffold has 1004 bp. The tenth percentile for size is 1,563 and the ninetieth is 56,612.

Let's plot the fraction of the scaffold, that is, N as a function of its size.

```
%matplotlib inline
import matplotlib.pyplot as plt


small_split =4800
large_split =540000
fig, axs = plt.subplots(1, 3, figsize=(16, 9), squeeze=False)
xs, ys = zip(*[(x, 100* y) for x, y in size_N if x <= small_split])
axs[0, 0].plot(xs, ys, '.')
axs[0, 0].set_ylim(-0.1, 3.5)
xs, ys = zip(*[(x, 100* y) for x, y in size_N if x > small_split and x <=
large_split])
axs[0, 1].plot(xs, ys, '.')
axs[0, 1].set_xlim(small_split, large_split)
xs, ys = zip(*[(x, 100* y) for x, y in size_N if x > large_split])
axs[0, 2].plot(xs, ys, '.')
axs[0, 0].set_ylabel('Fraction of Ns', fontsize=12)
axs[0, 1].set_xlabel('Contig size', fontsize=12)
fig.suptitle('Fraction of Ns per contig size', fontsize=26)
<matplotlib.text.Text at 0x113f039e8>
```

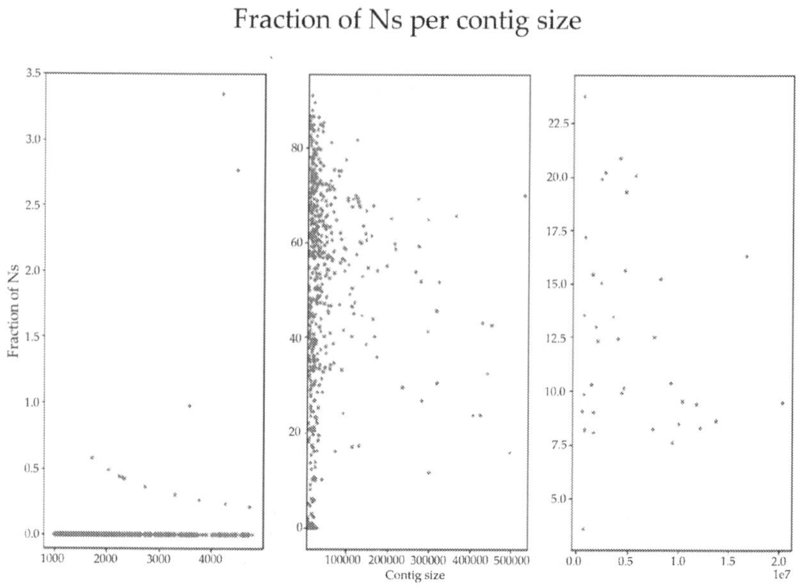

The plot is split into three parts based on scaffold size: one for scaffolds with less than 4,800 bp, one for scaffolds between 4,800 and 540,000 bp, and one for larger ones. Note that the y axis scale (depicting the fraction of the N scaffold) is completely different for the three panels. It is very low for small scaffolds (always below 44.5) with large variance (between 0 and 100 percent), for medium scaffolds and tighter variance (between 0 and 25 percent) for the largest scaffold.

The gambiae genome was *soft-masked*, meaning that some computational runs were done for identifying low-complexity regions in the genome. The repeats can also be identified using tools like *RepeatMasker*.

GENOME ANNOTATIONS TRAVERSAL

Having a genome sequence is interesting, but we will want to extract features from it: genes, exons, and coding sequences. This type of annotation information is made available in GFF and GTF. GFF stands for **Generic Feature Format**.

We will use the *gffutils* library to process the annotation file. Please install it before you proceed.

You need to create an annotation database with *gffutils* based on the GFF file.

Before that you need to download the following files to you disk.

http://www.vectorbase.org/download/anopheles-gambiae-pestbasefeaturesagamp42gtfgz

This will download a GTF file. Now, copy and paste the following link in your browser

http://www.vectorbase.org/download/anopheles-gambiae-pestbasefeaturesagamp42gff3gz

This should download a GFF file. (You may get gff3 extension file. Don't worry about it, just proceed). There are many variations of the GFF/GTF format. There are different GFF versions and many unoffcial variations. If possible, choose the GFF Version 3. However, the ugly truth is that you will find it very difficult to process files. The *gffutils* library tries as best as it can to accommodate this. Indeed, much of the documentation of the library is concerned with helping you process all kinds of awkward variations.

```
import gffutils
import sqlite3
try:
    db = gffutils.create_db('gambiae.gff3', 'ag.db')
except sqlite3.OperationalError:
    db = gffutils.FeatureDB('ag.db')
```

The gffutils library creates a SQLite database to store annotations efficiently. Here, we try to create the database, but if it already exists, we will use the existing one. This step can be time-consuming.

We now list all the available feature types and count them.

```
print(list(db.featuretypes()))

for feat_type in db.featuretypes():

   print(feat_type, db.count_features_of_type(feat_type))

['CDS', 'RNase_P_RNA', 'SRP_RNA', 'contig', 'exon', 'five_prime_UTR', 'gene', 'mRNA', 'miRNA', 'misc_RNA', 'pseudogene', 'rRNA', 'snRNA', 'snoRNA', 'tRNA', 'tRNA_pseudogene', 'three_prime_UTR']

CDS 62408

RNase_P_RNA 1

SRP_RNA 3

contig 8

exon 66485

five_prime_UTR 10520

gene 13624

mRNA 14697

miRNA 187

misc_RNA 10

pseudogene 5

rRNA 53

snRNA 38

snoRNA 12

tRNA 463

tRNA_pseudogene 9

three_prime_UTR 7281
```

Features will include contigs, genes, exons, transcripts, and so on. Note that we will use the gffutils package's featuretypes function. It will return a generator, but we will convert it to a list (it's safe here).

List all contigs here by doing following.

```
for contig in db.features_of_type('contig'):
  print(contig)
```

2L VectorBase contig 1 49364325 . . . ID=2L;molecule_type=dsDNA;translation_table=1;topology=linear;localization=chromosomal

3R VectorBase contig 1 53200684 . . . ID=3R;molecule_type=dsDNA;translation_table=1;topology=linear;localization=chromosomal

UNKN VectorBase contig 1 42389979 . . . ID=UNKN;molecule_type=dsDNA;translation_table=1;topology=linear;localization=chromosomal

X VectorBase contig 1 24393108 . . . ID=X;molecule_type=dsDNA;translation_table=1;topology=linear;localization=chromosomal

Y_unplaced VectorBase contig 1 237045 . . . ID=Y_unplaced;molecule_type=dsDNA;translation_table=1;topology=linear;localization=chromosomal

Mt VectorBase contig 1 15363 . . . ID=Mt;molecule_type=dsDNA;translation_table=1;topology=linear;localization=chromosomal

2R VectorBase contig 1 61545105 . . . ID=2R;molecule_type=dsDNA;translation_table=1;topology=linear;localization=chromosomal

3L VectorBase contig 1 41963435 . . . ID=3L;molecule_type=dsDNA;translation_table=1;topology=linear;localization=chromosomal

It appears that the genome has annotations for all the chromosome arms and sex chromosomes etc. We now venture into extracting useful information per chromosome among other things.

```
from collections import defaultdict
num_mRNAs = defaultdict(int)
num_exons = defaultdict(int)
max_exons =0
```

```
max_span =0
for contig in db.features_of_type('contig'):
   cnt =0
for gene in db.region((contig.seqid, contig.start, contig.end), featuretype='gene'):
      cnt +=1
      span = abs(gene.start - gene.end) # strand
if span > max_span:
         max_span = span
         max_span_gene = gene
      my_mRNAs = list(db.children(gene, featuretype='mRNA'))
      num_mRNAs[len(my_mRNAs)] +=1
if len(my_mRNAs) ==0: # checking alternative transcripts
         exon_check = [gene]
else:
         exon_check = my_mRNAs
for check in exon_check:
         my_exons = list(db.children(check, featuretype='exon'))
         num_exons[len(my_exons)] +=1
if len(my_exons) > max_exons:
            max_exons = len(my_exons)
            max_exons_gene = gene
   print('contig %s, number of genes %d'% (contig.seqid, cnt))
print('Max number of exons: %s (%d)'% (max_exons_gene.id, max_exons))
print('Max span: %s (%d)'% (max_span_gene.id, max_span))
print(num_mRNAs)
print(num_exons)
contig 2L, number of genes 3105
contig 3R, number of genes 2763
contig UNKN, number of genes 567
```

contig X, number of genes 1097

contig Y_unplaced, number of genes 0

contig Mt, number of genes 37

contig 2R, number of genes 3834

contig 3L, number of genes 2221

Max number of exons: AGAP001660 (67)

Max span: AGAP006656 (365621)

defaultdict(<class 'int'>, {2: 910, 1: 11595, 3: 211, 4: 74, 0: 781, 11: 3, 5: 27, 8: 4, 12: 1, 7: 5, 6: 9, 13: 1, 10: 1, 20: 1, 9: 1})

defaultdict(<class 'int'>, {4: 2091, 2: 3359, 5: 1411, 6: 1039, 1: 2019, 3: 2838, 9: 419, 10: 298, 11: 202, 8: 454, 12: 159, 31: 5, 7: 718, 13: 106, 15: 65, 19: 28, 16: 45, 17: 53, 14: 65, 26: 3, 18: 22, 21: 9, 22: 7, 24: 6, 30: 5, 20: 19, 32: 1, 33: 1, 27: 2, 28: 5, 23: 6, 34: 1, 29: 4, 25: 9, 67: 1, 50: 1, 49: 1, 42: 1})

We traverse all contigs (features_of_type), extracting all genes (region). In each gene, we count the number of alternative transcripts. If there are none (note that this is probably an annotation issue and not a biological one), we count the exons (children). If there are several transcripts, we count the exons per transcript. We also account for the span size to check for the gene spanning the largest region. We follow a similar procedure to find the gene and the largest number of exons.

we print a dictionary with the distribution of the number of alternative transcripts per gene (*num_mRNAs*) and the distribution of number of exons per transcript (*num_exons*).

EXTRACTING INFORMATION ABOUT A SPECIFIC GENE

Let us now venture into extracting information about a gene, from a reference, using annotations. We aim to extract a gene sequence with the help of an annotation file to get its coordinates against a reference FASTA. We will use the Anopheles gambiae genome, along with its annotation file for achieving this objective. The gene of interest is a **Voltage-gated sodium channel** (VGSC), involved in resistance to pesticides.

```
gene_id ='AGAP004707'
gene = db[gene_id]
print(gene)
print(gene.seqid, gene.strand)
2L	VectorBase	gene	2358158	2431617	.	+	.
ID=AGAP004707;biotype=protein_coding
2L +
```

You need to gene ID before you venture into this type of data extraction. We knew our gene ID from an online database called VectorBase. You can see that the gene is on the 2L chromosome arm and coded in the positive direction.

```
recs = SeqIO.parse('gambiae.fa', 'fasta')
for rec in recs:
   print(rec.description)
if rec.description.split(':')[2] == gene.seqid:
      my_seq = rec.seq
break
print(my_seq.alphabet)
chromosome:AgamP3:2L:1:49364325:1 chromosome 2L
SingleLetterAlphabet()
```

Create a function to construct a gene sequence for a list of CDSs.

```
from Bio import Alphabet, Seq, SeqIO
def get_sequence(chrom_seq, CDSs, strand):
   seq = Seq.Seq('', alphabet=Alphabet.IUPAC.unambiguous_dna)
for CDS in CDSs:
          my_cds = Seq.Seq(str(my_seq[CDS.start -1: CDS.end]),
alphabet=Alphabet.IUPAC.unambiguous_dna)
      seq += my_cds
return seq if strand =='+'else seq.reverse_complement()
```

This function will receive a chromosome sequence (in our case, the 2L arm), a list of coding sequences (retrieved from the annotation file), and the strand. We will construct a sequence of the unambiguous DNA type (we will need this for translation). The Single Letter Alphabet of the previous step will have to be converted. We also have to be very careful with the start and end of the sequence (note that the GFF file is 1-based, whereas the Python array is 0-based). Finally, we return the reverse complement if the strand is negative.

```
# choose one transcript of the three available for this gene ID
mRNAs = db.children(gene, featuretype='mRNA')
for mRNA in mRNAs:
    print(mRNA.id)
if mRNA.id.endswith('RA'):
break
# get the coding sequence of the transcript
CDSs = db.children(mRNA, featuretype='CDS', order_by='start')
gene_seq = get_sequence(my_seq, CDSs, gene.strand)
print(len(gene_seq), gene_seq)
prot = gene_seq.translate()
print(len(prot), prot)
AGAP004707-RA
6357
```

ATGACCGAAGACTCCGATTCGATATCTGAGGAAGAACG-
TAGTTTGTTCCGTCCTTTCACTCGTGAATCATTACAAGCTATC-
GAAGCACGCATTGCAGATGAAGAAGCCAAACAGCGAGA-
ATTGGAAAGAAAACGAGCTGAGGGGGAGGATGAGGAT-
GAAGGTCCCCAACCGGACCCTACTCTTGAACAGGGTGTAC-
CAGTCCCAGTTCGAATGCAAGGCAGCTTCCCCCCGGAGTTG-
GCCTCCACGCCTCTCGAGGATATTGACAGTTTCTATTCAAAT-
CAAAGGACATTCGTAGTGATTAGTAAAGGAAAAGATATATTTC-
GTTTCTCCGCAACTAACGCATTATATGTACTTGATCCGTTTA-
ACCCCATACGCCGCGTAGCTATTTATATTTTAGTACATCCACT-
GTTTTCACTTTTTATAATAACGACCATTCTTGTTAATTGTATATT-
GATGATTATGCCTACCACGCCGACAGTCGAATCTACCGAGGT-
GATATTCACCGGCATCTACACGTTCGAATCAGCTGTAAAAGT-
GATGGCGCGAGGTTTCATATTACAACCGTTTACTTATCTTAGA-

GATGCATGGAATTGGTTGGACTTCGTAGTAATAGCATTAG-
CATATGTAACTATGGGTATAGATTTGGGTAATCTTGCTGCGTT-
GAGAACATTCAGGGTATTACGAGCTCTTAAAACGGTAGC-
CATCGTTCCAGGCTTAAAAACCATCGTCGGAGCCGT-
TATAGAATCCGTAAAGAATCTCAGAGATGTGATAATTTTAA-
CAATGTTTTCGTTATCCGTGTTTGCTTTGATGGGTCTA-
CAAATCTACATGGGAGTACTAACACAAAAGTGCATA-
AAAGAGTTCCCATTGGATGGTTCCTGGGGTAATCTAACCGAC-
GAAAGCTGGGAGCTGTTCAACAGCAATGACACAAATTG-
GTTCTATTCCGAGAGTGGCGACATTCCTCTTTGTGGAAACT-
CATCTGGAGCTGGACAATGTGATGAAGGCTACATTTGTTTA-
CAAGGCTATGGCAAAAATCCAAATTACGGGTATACAAGTTTT-
GATACATTCGGATGGGCATTCTTGTCTGCCTTTCGTCTAAT-
GACTCAAGATTATTGGGAGAATTTATATCAACTGGTGTTAC-
GATCAGCTGGACCGTGGCACATGCTCTTCTTCATTGTGAT-
TATCTTCTTGGGTTCGTTTTACCTTGTAAATTTAATCTTGGC-
CATTGTCGCCATGTCGTACGACGAACTCCAGAAGAAGGCC-
GAAGAGGAAGAGGCCGCCGAGGAAGAAGCACTTC-
GGGAAGCAGAAGAAGCTGCGGCAGCAAAGGCAGCTA-
AACTGGAAGCACAACAAGCGGCCGCAGCAGCAGCGGC-
GAACCCAGAAATCGCTAAAAGCCCTTCGGATTTCTCCTGTCA-
CAGCTATGAGTTGTTTGTCGGACAGGAGAAAGGCAACGAC-
GATAACAATAAGGAGAAAATGTCCATTAGAAGCGAAGGATTG-
GAGTCGGTGAGCGAAATCACAAGAACAACCGCACCAACAGC-
TACTGCAGCTGGCACTGCAAAAGCCCGTAAAGTGAGCGCG-
GCTTCACTTTCATTACCCGGTTCACCATTTAATCTTCGTAGAG-
GATCTAGAGGATCACACCAGTTCACGATACGTAACGGTAGAG-
GACGTTTCGTTGGTGTACCTGGTAGCGATAGAAAACCACTGG-
TACTATCAACATATCTCGATGCACAAGAACACCTGCCATAC-
GCTGATGATTCCAACGCGGTAACGCCGATGTCG-
GAAGAAAATGGTGCAATCATCGTTCCAGTATACTATGCTAATT-
TAGGTTCAAGGCACTCGTCGTATACTTCGCATCAGTCGCG-
TATTTCGTACACATCTCACGGTGACCTGCTCGGGGGCAT-
GACAAAAGAGAGCCGTCTGCGAAATCGATCAGCCCGTAA-
CACTAACCATTCAATTGTACCACCTCCGAATGCTAACAATC-
TATCCTACGCTGAAACAAACCATAAAGGACAGCGAGATTTC-
GACTTGACACAGGACTGTACAGACGATGCCGGAAAAATA-
AAACATAACGACAATCCTTTCATAGAACCTGCTCAAACT-

CAAACTGTGGTAGATATGAAAGACGTGATGGTGTTAAAT-
GACATTATTGAGCAAGCTGCTGGTCGGCACAG-
CAGGGCAAGTGATCACGGAGTCTCTGTTTATTACTTCCCCA-
CAGAGGACGACGACGAAGATGGCCCAACGTTTAAGGA-
CAAAGCCCTCGAGTTTCTGATGAAGATGATCGACATTTTCTGT-
GTGTGGGACTGTTGTTGGGTTTGGCTTAAATTTCAG-
GAATGGGTTGCTTTTATTGTGTTTGACCCATTTGTAGAGC-
TATTCATTACGCTCTGCATTGTGGTAAATACACTGTTTATG-
GCTTTAGATCATCACGATATGGATCCAGATATG-
GAAAAGGCACTGAAAAGTGGCAACTACTTCTTCACAGCTA-
CATTTGCCATCGAAGCTACAATGAAGCTCATAGCAAT-
GAGCCCCAAATATTACTTTCAAGAGGGTTGGAATATCTTC-
GATTTTATTATCGTAGCACTGTCTCTGCTAGAATTGGGACTT-
GAGGGTGTTCAAGGATTGTCAGTATTACGATCGTTCCGTTTGC-
TAAGAGTTTTCAAGCTGGCAAAATCATGGCCTACATT-
GAATCTTCTAATTTCCATCATGGGACGTACAGTTGGT-
GCCCTTGGTAATTTAACCTTCGTCTTATGCATCAT-
CATTTTCATCTTCGCCGTGATGGGGATGCAACTTTTTG-
GCAAAAACTACACAGATAATGTGGATAGATTCCCCGACCAT-
GATCTGCCAAGATGGAATTTTACAGATTTCATG-
CATTCCTTCATGATTGTGTTCCGTGTGCTATGCGGAGAATG-
GATTGAATCAATGTGGGATTGTATGCTTGTCGGTGATG-
TATCCTGCATACCATTTTTCTTGGCCACTGTAGTGATAG-
GAAATTTAGTCGTGCTTAACCTTTTCTTAGCCTTGCTTTTGT-
CAAATTTTGGTTCATCATCCTTGTCTGCACCAACGGCAGATA-
ATGAGACCAACAAGATTGCAGAAGCGTTCAACAGAATATCAC-
GCTTTTCTAACTGGATTAAAATGAATTTAGCAAACGCTCT-
CAAGTTTGTAAAAAATAAATTAACAAGCCAAATAGCATCC-
GTTCAACCGACAGGCAAAGGCGTATGTCCATG-
TATATCTTCAGAGCATGGTGAAAATGAACTGGAACTTACTC-
CAGACGATATTTTGGCGGATGGACTATTGAAGAAAGGAAT-
CAAAGAGCACAACCAACTGGAAGTAGCGATTGGCGATG-
GCATGGAATTCACCATTCATGGTGATCTGAAGAA-
CAAAGCAAAAAAGAATAAACAAATCATGAACAACTCTA-
AGGTGATAGGCAATTCTATTAGTAATCATCAAGATAATAAATTA-
GATCATGAACTGAATCATAGAGGCGTGTCCTTACAGGACGAT-
GATACTGCTAGTATCAAATCTTATGGCAGTCACAAGAATC-
GCCCATTTAAGGATGAAAGCCACAAAGGCAGCGCCGAAAC-

GATGGAGGGTGAAGAAAAACGTGATGCCAGCAAGGAGGATC-
TAGGAATTGACGAAGAACTCGACGACGAAGGCGAGGGAGAT-
GAAGGTCCTCTGGATGGAGAGCTGATTATTCATG-
CAGAAGAAGACGAAGTGATTGAGGATTCACCGGCGGATT-
GCTGCCCGGACAACTGCTACAAAAAATTTCCTGTTCTT-
GCTGGGGATGATGACGCGCCGTTCTG-
GCAGGGTTGGGGAAATTTACGTCTCAAAACGTTTCAGCTA-
ATAGAAAATAAGTATTTTGAGACAGCTGTAATTACAAT-
GATTCTGCTTAGTAGCTTAGCTTTGGCCCTCGAAGATGTG-
CATCTTCCACAGCGCCCAATCCTTCAAGATATTCTTTATTA-
CATGGATCGAATTTTCACAGTGATCttttttttAGAGATGTTAAT-
CAAATGGTTAGCTTTAGGTTTTAAAGTATATTTTACAAAT-
GCTTGGTGTTGGCTTGATTTCATTATCGTGATGGTATCTTT-
GATAAACTTCGTTGCTTCACTTTGTGGAGCTGGTGG-
TATTCAAGCATTCAAAACCATGCGAACTCTTAGAGCCCT-
GAGACCACTACGTGCCATGTCCCGTATGCAGGGAAT-
GAGGGTCGTCGTGAATGCGTTGGTTCAAGCTATACCGTC-
CATCTTCAACGTGCTGCTGGTTTGTTTGATATTCTGGCTA-
ATATTTGCAATCATGGGGGTGCAATTATTTGCTG-
GCAAATACTTCAAGTGTGTGGATAAAAATAAAACTACAT-
TACCTCACGAAATTATACCGGATGTAAATGCTTGCAAAGCC-
GAAAACTATTCATGGGAAAATTCACCAATGAACTTCGATCAT-
GTAGGTAAAGCATATTTGTGTCTGTTTCAAGTAGCCA-
CATTCAAAGGATGGATACAAATAATGAACGATGCTATTGATTC-
TAGAGACGTAAGTTTTGTCGGTAAACAGCCTATACGGGAAAC-
GAATATCTACATGTATCTGTACTTTGTGTTCTTTAT-
TATCTTTGGGTCATTCTTCACGTTGAATCTATTCATTGGTGT-
TATAATTGACAACTTCAATGAACAGAAAAAGAAAGCTGGTG-
GATCGCTAGAAATGTTCATGACAGAGGATCAGAAAAAGTAC-
TATAATGCAATGAAAAAAATGGGTTCGAAGAAACCTCTA-
AAGGCAATTCCTCGTCCAAGGTGGCGGCCT-
CAAGCAATAGTTTTTGAAATAGTGACGAACAAAAAGTTT-
GACATGATTATCATGTTGTTCATCGGATTCAATATGTTAACTAT-
GACACTGGACCACTACAAACAATCAGAAACTTTTAGTGCT-
GTTTTGGATTACTTGAATATGATATTCATCTGCATATTCAG-
CAGCGAATGTTTAATGAAGATTTTTGCACTTCGTTATCAT-
TACTTTATCGAGCCATGGAATTTGTTTGATTTTGTTGTCGT-
CATTCTTTCGATTTTGGGCCTTGTTCTAAGTGATATCATT-

GAAAAATATTTTGTATCTCCCACACTTCTACGAGTCGTGC-
GAGTGGCAAAAGTGGGCCGAGTATTGCGTTTGGTTA-
AAGGAGCCAAGGGTATCCGAACGTTGCTGTTTGCATTAG-
CAATGTCGCTACCTGCACTATTTAACATCTGCTTGT-
TACTCTTTTTGGTGATGTTTATATTTGCCATTTTTGGAATGT-
CATTTTTCATGCACGTCAAAGATAAGAGTGGCTTAGATGAC-
GTGTACAATTTTAAAACGTTTGGCCAGAGCATGATTTTAC-
TATTTCAGATGTCAACCTCGGCTGGGTGGGATGGTGTTTTAG-
ATGGTATTATCAATGAAGAAGACTGTCTTCCACCAGACAAT-
GATAAGGGCTATCCGGGAAATTGTGGTTCATCAACAATTG-
GCATAACGTACTTATTGGCGTATCTTGTAATAAGTTTCCTTATC-
GTTATTAACATGTACATTGCTGTTATCCTCGAAAACTACTCG-
CAAGCTACGGAAGATGTTCAAGAAGGCTTAACTGATGACGAT-
TATGATATGTACTACGAAATATGGCAGCAATTCGATCCTGACG-
GTACACAATACGTTCGATATGATCAGCTATCAGACTTTTTG-
GATGTGCTGGAACCGCCTCTACAGATTCATAAACCAAATCGT-
TATAAGATTATTTCGATGGATATTCCGATATGCCGCGGAGATAT-
GATGTTCTGTGTCGATATTCTAGATGCACTAACGAAA-
GATTTTTTTGCTAGAAAAGGAAATCCTATAGAAGAAACAGCC-
GAATTAGGTGAAGTTCAACAACGCCCAGACGAAGTTGGT-
TACGAACCAGTATCATCAACACTTTGGAGGCAGCGTGAAGAG-
TACTGTGCTCGACTGATACAGCATGCGTGGAAACGCTATAAA-
CAGCGTCACGGAGGCGGAACAGACGCTTCAGGAGAT-
GATCTTGAAATAGATGCCTGTGATAACGGTTGTGGTGGTGG-
TAATGGCAATGAAAATGATGATAGTGGAGATGGTGCAACAGG-
TAGTGGTGACAACGGAAGTCAGCATGGTGGTGGCAGCATA-
AGTGGCGGAGGAGGAACTCCTGGTGGTGGTAAAAGTA-
AAGGAATTATTGGCAGTACTCAGGCTAACATAGGCATAGTG-
GATAGTAATATATCACCAAAGGAATCACCGGATAGCATCGGC-
GATCCCCAAGGTCGTCAGACGGCCGTTCTTGTGGAGAGC-
GACGGATTTGTGACGAAAAACGGTCACCGTGTCGTCATA-
CACTCTCGATCTCCCAGCATAACATCGCGAACGGCAGAT-
G T C T G A

2119 MTEDSDSISEEERSLFRPFTRESLQAIEARIADEEAK-
QRELERKRAEGEDEDEGPQPDPTLEQGVPVPVRMQGSFPPE-
LASTPLEDIDSFYSNQRTFVVISKGKDIFRFSATNALYVLDPFN-
PIRRVAIYILVHPLFSLFIITTILVNCILMIMPTTPTVESTEVIFTGI-
YTFESAVKVMARGFILQPFTYLRDAWNWLDFVVIALAYVT-

MGIDLGNLAALRTFRVLRALKTVAIVPGLKTIVGAVIES-
VKNLRDVIILTMFSLSVFALMGLQIYMGVLTQKCIKEFPLDGSW-
GNLTDESWELFNSNDTNWFYSESGDIPLCGNSSGAGQCDEGY-
ICLQGYGKNPNYGYTSFDTFGWAFLSAFRLMTQDYWENLY-
QLVLRSAGPWHMLFFIVIIFLGSFYLVNLILAIVAMSYDELQK-
KAEEEEAAEEEALREAEEAAAAKAAKLEAQQAAAAAAAN-
PEIAKSPSDFSCHSYELFVGQEKGNDDNNKEKMSIRSEGLESV-
SEITRTTAPTATAAGTAKARKVSAASLSLPGSPFNLRRGSRG-
SHQFTIRNGRGRFVGVPGSDRKPLVLSTYLDAQEHL-
PYADDSNAVTPMSEENGAIIVPVYYANLGSRHSSYTSHQSRISYT-
SHGDLLGGMTKESRLRNRSARNTNHSIVPPPNANNLSYAETNH-
KGQRDFDLTQDCTDDAGKIKHNDNPFIEPAQTQTVVDMKDVM-
VLNDIIEQAAGRHSRASDHGVSVYYFPTEDDDEDGPTFKD-
KALEFLMKMIDIFCVWDCCWVWLKFQEWVAFIVFDP-
FVELFITLCIVVNTLFMALDHHDMDPDMEKALKSGNYFFTAT-
FAIEATMKLIAMSPKYYFQEGWNIFDFIIVALSLLELG-
LEGVQGLSVLRSFRLLRVFKLAKSWPTLNLLISIMGRTVGALGN-
LTFVLCIIIFIFAVMGMQLFGKNYTDNVDRFPDHDLPRWNFTD-
FMHSFMIVFRVLCGEWIESMWDCMLVGDVSCIPFFLATVVI-
GNLVVLNLFLALLLSNFGSSSLSAPTADNETNKIAEAFNRISRFS-
NWIKMNLANALKFVKNKLTSQIASVQPTGKGVCPCISSEH-
GENELELTPDDILADGLLKKGIKEHNQLEVAIGDGMEFTIHG-
DLKNKAKKNKQIMNNSKVIGNSISNHQDNKLDHELNHRGVS-
LQDDDTASIKSYGSHKNRPFKDESHKGSAET-
MEGEEKRDASKEDLGIDEELDDEGEGDEGPLDGELIIHAEEDEV-
IEDSPADCCPDNCYKKFPVLAGDDDAPFWQGWGNLRLKT-
FQLIENKYFETAVITMILLSSLALALEDVHLPQRPILQDILYYM-
DRIFTVIFFLEMLIKWLALGFKVYFTNAWCWLDFIIVMVSLIN-
FVASLCGAGGIQAFKTMRTLRALRPLRAMSRMQGMRV-
VVNALVQAIPSIFNVLLVCLIFWLIFAIMGVQLFAGKYFKCVD-
KNKTTLPHEIIPDVNACKAENYSWENSPMNFDHVGKAYLCLFQ-
VATFKGWIQIMNDAIDSRDVSFVGKQPIRETNIYMYLYFVFFI-
IFGSFFTLNLFIGVIIDNFNEQKKKAGGSLEMFMTEDQKKYY-
NAMKKMGSKKPLKAIPRPRWRPQAIVFEIVTNKKFDMIIMLFIG-
FNMLTMTLDHYKQSETFSAVLDYLNMIFICIFSSECLMKIFALRY-
HYFIEPWNLFDFVVVILSILGLVLSDIIEKYFVSPTLLRVVR-
VAKVGRVLRLVKGAKGIRTLLFALAMSLPALFNICLLLFLVMFI-
FAIFGMSFFMHVKDKSGLDDVYNFKTFGQSMILLFQMSTSAG-

WDGVLDGIINEEDCLPPDNDKGYPGNCGSSTIGITYLLAYLV-ISFLIVINMYIAVILENYSQATEDVQEGLTDDDYDMYYEI-WQQFDPDGTQYVRYDQLSDFLDVLEPPLQIHKPNRYKIISMDIPI-CRGDMMFCVDILDALTKDFFARKGNPIEETAELGEVQQRPDEV-GYEPVSSTLWRQREEYCARLIQHAWKRYKQRHGGGTDAS-GDDLEIDACDNGCGGGNGNENDDSGDGATGSGDNGSQHGGG-SISGGGGTPGGGKSKGIIGSTQANIGIVDSNISPKESPDSIG-DPQGRQTAVLVESDGFVTKNGHRVVIHSRSPSITSRTADV*

Let's get the gene that is coded in the negative strand direction. We will just take the gene next to VGSC (which happens to be the negative strand):

```
reverse_transcript_id ='AGAP004708-RA'
reverse_CDSs = db.children(reverse_transcript_id, featuretype='CDS', order_by='start')
reverse_seq = get_sequence(my_seq, reverse_CDSs, '-')
print(len(reverse_seq), reverse_seq)
reverse_prot = reverse_seq.translate()
print(len(reverse_prot), reverse_prot)
```

1992 ATGGCTGACTTCGATAGTGCCACTAAATGTATCAGAAA-CATTGAAAAAGAAATTCTTCTCTTGCAATCCGAAGTTTT-GAAGACTCGTGAGGGGCTTGGGCTGGAAGATGATAAC-GTGGAACTTAAAAAGTTAATGGAGGAAAACACGAGAT-TAAAGCATCGTTTGGAGATAGTGCAATCGGCTATTGTA-CAGGAAGGCGGATCAATCGCATCCTCCGATTCTGGCAAC-CAATCCATTGTTGGCGAACTGCAGCAAGTATTTACCGAAGC-CATTCAAAAAGCTTTTCCAAGTGTGTTGGTTGAGGCGGT-TATTACTATTTCGTCATCCCCCAAGTTTGGCGATTATCAAT-GCAATAGTGCTATGCAGATTGCGCAGCATTTGAAGCAGT-TATCTGTTAAATCGTCGCCACGTGAAGTGGCCCAAAAACTG-GTAGCTGAATTGCAAAAACCAATACCTTGTGTCGATAG-ATTAGAAATCGCTGGAGCGGGATACGTTAATATTTTCCTGTC-TAGATCTTATGGAGAACAACGCATTATGAGCATCTTGAG-GCATGGGATTGTGGTACCATTAATAGAAAAGAAACGTGT-GATAGTCGATTTTTCCTCGCCTAACGTAGCGAAAGAAATGCAT-GTCGGTCATTTACGTTCGACCATCATTGGTGATTCAATTTGTC-

GATTTTTGGAATATCTCGGACACGATGTGCTTCGTATTAAC-
CATATCGGAGACTGGGGAACGCAATTTGGTATGTTAATTGCT-
CATTTGCAGGACCGTTTCCCTAATTTCCAAACCGAGTCCCC-
GCCTATCAGCGATTTGCAAGCATTTTACAAGGAGT-
CAAAGGTCCGATTTGACAGCGATGAAGTATTTAAAAAGC-
GTGCCTACGAATGTGTAGTCAAACTGCAAAGTGGAGAGCT-
GAGTTATTTGAAGGCCTGGAATCTAATTTGCGATGTTTCAC-
GCAAAGAATTCCAAACCATCTACAACAGATTGGATGTGAAAC-
TAGTTGAACGTGGTGAATCGTTTTATCAAAGCAGAATG-
GAAAAAATCGTAGAAGAACTTAAGCAGGATGGGTTCCTT-
GAAGAAGACGAAGGCCGTCTTATCATGTGGGGCGAAAATC-
GCGCTGGAATTCCTTTAACCATCGTAAAATCAGACGGAG-
GATTTACATATGATACTTCGGATATGGCCGCCATCAAACAAC-
GCTTGCAAGAAGAAAAGGCTGATTGGTTGATATATGTAACT-
GACGCTGGGCAGGCGACTCATTTCCAAACAATTTTTCTT-
GTGCAAAACGAGCCAAAATCCTACAAGAGAGCAAACATC-
GTGTGGATCACGTCGGATTTGGTGTGGTGCTAGGCGAA-
GATGGTAAAAAATTCAAGACTCGTTCTGGCGATACGGT-
GAAATTGACAGAACTTCTCAATGAAGGTTTGAGGAGGGC-
TATGGAAAAACTAGTTCAGAAGGAAAGGAACTTAGTGCT-
CACACAAGAGGAGCTAGTTGCAGCACAAGAATCAGTC-
GCCTACGGTTGTATTAAATATGCGGATCTGTCGCATAATCG-
TAACAACGAATATGTGTTCTCCTTCGATAAGATGCTGGAGGA-
CAAAGGAAATACTGCCGTGTATCTGTTGTATGCCTATACCC-
GCATTCGTTCTATTGCAAGAAAATGTGGCGGAGATTTTG-
CAAATGACATGCAAAAGGTGATCGATTCCACAGTTATTA-
AATTAGATCATGAAAAGGAATGGAAACTCGCCAAGGTGTT-
GCTTCGTTTTACCGACGTTATGTTATTGATCATGAAAAATC-
TATCGTTACATCATCTTTGTGAGTTTGTGTACGAAATATG-
CACTGCTTTTAGTGAGTTTTATGACAGTTGTTATTGCATC-
GAAAAAAATAAGCAAGGTGAAATTATTACTGTTTATCCCTCTC-
GCGTCTTGCTATGCGAAGCAACATCAAAGGTGCTGGAAAAAT-
GTTTCGATATTTTAGGACTGAAGCCTGTGCATAAAATATAA

664 MADFDSATKCIRNIEKEILLLQSEVLKTREGLGLEDDN-
VELKKLMEENTRLKHRLEIVQSAIVQEGGSIASSDSGNQ-
SIVGELQQVFTEAIQKAFPSVLVEAVITISSSPKFGDYQCNSAM-
QIAQHLKQLSVKSSPREVAQKLVAELQKPIPCVDRLEIAGA-
GYVNIFLSRSYGEQRIMSILRHGIVVPLIEKKRVIVDFSSPNVAKE-

MHVGHLRSTIIGDSICRFLEYLGHDVLRINHIGDWGTQFGM-LIAHLQDRFPNFQTESPPISDLQAFYKESKVRFDSDEVFK-KRAYECVVKLQSGELSYLKAWNLICDVSRKEFQTIYNRLD-VKLVERGESFYQSRMEKIVEELKQDGFLEEDEGRLIMWGEN-RAGIPLTIVKSDGGFTYDTSDMAAIKQRLQEEKADWLIYVT-DAGQATHFQTIFSCAKRAKILQESKHRVDHVGFGVVLGEDGK-KFKTRSGDTVKLTELLNEGLRRAMEKLVQKERNLVLTQEEL-VAAQESVAYGCIKYADLSHNRNNEYVFSFDKMLEDKGN-TAVYLLYAYTRIRSIARKCGGDFANDMQKVIDSTVIKLDHEK-EWKLAKVLLRFTDVMLLIMKNLSLHHLCEFVYEICTAFSEFYD-SCYCIEKNKQGEIITVYPSRVLLCEATSKVLEKCFDILGLKPVH-KI*

FINDING ORTHOLOGS

We now venture into finding orthologs of a gene of interest. Although web services are available for doing this, we will develop a recipe which can be utilized for regular tasks. It can be tailored for your specific jobs easily. This introduces the reader to retrieving orthologs using REST APIs on the Web. Also, one gets to learn to access programmatically **Ensembl** database.

```
import requests
ensembl_server ='http://rest.ensembl.org'


def do_request(server, service, *args, **kwargs):
   url_params =''
for a in args:
         if a is not None:
                        url_params +='/'+ a
         req = requests.get('%s/%s%s'% (server, service, url_params),
              params=kwargs,
              headers={'Content-Type': 'application/json'})
if not  req.ok:
```

```
    req.raise_for_status()
return req.json()
```

We start by importing the *requests* library and specifying the root URL. Then, we create a simple function that will take the functionality to be called (see the following examples) and generate a complete URL. It will also add optional parameters and specify the payload to be of the JSON type (just to get a default JSON answer). It will return the response in the JSON format. This is typically a nested Python data structure of lists and dictionaries.

```
answer = do_request(ensembl_server, 'info/species')
for sp in answer['species']:
  print(sp['name'])
saccharomyces_cerevisiae
ciona_savignyi
myotis_lucifugus
taeniopygia_guttata
sorex_araneus
otolemur_garnettii
macropus_eugenii
erinaceus_europaeus
anolis_carolinensis
gadus_morhua
dasypus_novemcinctus
chlorocebus_sabaeus
tursiops_truncatus
mus_musculus
bos_taurus
mus_musculus_wsbeij
monodelphis_domestica
choloepus_hoffmanni
sus_scrofa
```

rattus_norvegicus
caenorhabditis_elegans
pteropus_vampyrus
microcebus_murinus
sarcophilus_harrisii
ovis_aries
papio_anubis
mus_musculus_akrj
pelodiscus_sinensis
equus_caballus
xiphophorus_maculatus
macaca_mulatta
mus_musculus_balbcj
mus_musculus_dba2j
astyanax_mexicanus
latimeria_chalumnae
mus_musculus_nodshiltj
mus_musculus_aj
ficedula_albicollis
gasterosteus_aculeatus
oryctolagus_cuniculus
gorilla_gorilla
oreochromis_niloticus
echinops_telfairi
nomascus_leucogenys
homo_sapiens
dipodomys_ordii
mus_musculus_casteij
lepisosteus_oculatus
anas_platyrhynchos
canis_familiaris
callithrix_jacchus
pongo_abelii

ornithorhynchus_anatinus
tetraodon_nigroviridis
mustela_putorius_furo
mus_musculus_c57bl6nj
tarsius_syrichta
vicugna_pacos
meleagris_gallopavo
xenopus_tropicalis
mus_musculus_fvbnj
ictidomys_tridecemlineatus
mus_musculus_pwkphj
mus_musculus_nzohlltj
mus_musculus_129s1svimj
mus_musculus_cbaj
cavia_porcellus
takifugu_rubripes
ochotona_princeps
pan_troglodytes
petromyzon_marinus
ailuropoda_melanoleuca
mus_spretus_spreteij
felis_catus
mus_musculus_c3hhej
procavia_capensis
oryzias_latipes
danio_rerio
gallus_gallus
tupaia_belangeri
mus_musculus_lpj
ciona_intestinalis

loxodonta_africana

poecilia_formosa

drosophila_melanogaster

We will now try to nd any HGNC databases on the server related to human data:

```
ext_dbs = do_request(ensembl_server, 'info/external_dbs',
'homo_sapiens', filter='HGNC %')
print(ext_dbs)
```

[{'name': 'HGNC', 'release': '1', 'description': None, 'display_name': 'HGNC Symbol'}, {'name': 'HGNC_trans_name', 'release': '1', 'description': 'transcript name from HGNC', 'display_name': 'HGNC transcript name'}]

We restrict the search to human-related databases (*homo_sapiens*). We also filter databases starting with HGNC (this filtering uses the SQL notation). HGNC is the HUGO database. We want to make sure that it's available because the HUGO database is responsible for curating human gene names and maintains our LCT identifier.

Now that we know that the LCT identifier is probably available, we want to retrieve the Ensembl ID for the gene, as shown in the following code:

```
answer = do_request(ensembl_server, 'lookup/symbol', 'homo_sapiens',
'LCT')
print(answer)
lct_id = answer['id']
```

{'source': 'ensembl_havana', 'object_type': 'Gene', 'logic_name': 'ensembl_havana_gene', 'version': 9, 'species': 'homo_sapiens', 'description': 'lactase [Source:HGNC Symbol;Acc:HGNC:6530]', 'display_name': 'LCT', 'assembly_name': 'GRCh38', 'biotype': 'protein_coding', 'end': 135837180, 'seq_region_name': '2', 'db_type': 'core', 'strand': -1, 'id': 'ENSG00000115850', 'start': 135787840}

We can also inspect other databases that are known to Ensembl to refer to this gene:

```
lct_xrefs = do_request(ensembl_server, 'xrefs/id', lct_id)
```

```
for xref in lct_xrefs:
    print(xref[‘db_display_name’])
    print(xref)
```

Vega gene

{‘display_id’: ‘OTTHUMG00000131738’, ‘primary_id’: ‘OTTHUMG00000131738’, ‘version’: ‘2’, ‘description’: None, ‘dbname’: ‘Vega_gene’, ‘synonyms’: [], ‘info_type’: ‘NONE’, ‘info_text’: ‘Added during ensembl-vega production’, ‘db_display_name’: ‘Vega gene’}

Vega gene

{‘display_id’: ‘LCT’, ‘primary_id’: ‘OTTHUMG00000131738’, ‘version’: ‘1’, ‘description’: None, ‘dbname’: ‘Vega_gene’, ‘synonyms’: [], ‘info_type’: ‘NONE’, ‘info_text’: ‘’, ‘db_display_name’: ‘Vega gene’}

Havana gene

{‘display_id’: ‘OTTHUMG00000131738’, ‘primary_id’: ‘OTTHUMG00000131738’, ‘version’: ‘0’, ‘description’: None, ‘dbname’: ‘OTTG’, ‘synonyms’: [], ‘info_type’: ‘NONE’, ‘info_text’: ‘’, ‘db_display_name’: ‘Havana gene’}

Expression Atlas

{‘display_id’: ‘ENSG00000115850’, ‘primary_id’: ‘ENSG00000115850’, ‘version’: ‘0’, ‘description’: ‘’, ‘dbname’: ‘ArrayExpress’, ‘synonyms’: [], ‘info_type’: ‘DIRECT’, ‘info_text’: ‘’, ‘db_display_name’: ‘Expression Atlas’}

LRG display in Ensembl gene

{‘display_id’: ‘LRG_338’, ‘primary_id’: ‘LRG_338’, ‘version’: ‘0’, ‘description’: ‘Locus Reference Genomic record for LCT’, ‘dbname’: ‘ENS_LRG_gene’, ‘synonyms’: [], ‘info_type’: ‘DIRECT’, ‘info_text’: ‘’, ‘db_display_name’: ‘LRG display in Ensembl gene’}

NCBI gene

{‘display_id’: ‘LCT’, ‘primary_id’: ‘3938’, ‘version’: ‘0’, ‘description’: ‘lactase’, ‘dbname’: ‘EntrezGene’, ‘synonyms’: [], ‘info_type’: ‘DEPENDENT’, ‘info_text’: ‘’, ‘db_display_name’: ‘NCBI gene’}

HGNC Symbol

{'display_id': 'LCT', 'primary_id': 'HGNC:6530', 'version': '0', 'description': 'lactase', 'dbname': 'HGNC', 'synonyms': [], 'info_type': 'DIRECT', 'info_text': 'Generated via ensembl_manual', 'db_display_name': 'HGNC Symbol'}

MIM gene

{'display_id': 'LACTASE; LCT [*603202]', 'primary_id': '603202', 'version': '0', 'description': 'LACTASE; LCT;;LAC;;LACTASE-PHLORIZIN HYDROLASE; LPH', 'dbname': 'MIM_GENE', 'synonyms': [], 'info_type': 'DEPENDENT', 'info_text': '', 'db_display_name': 'MIM gene'}

MIM morbid

{'display_id': 'LACTASE DEFICIENCY, CONGENITAL [#223000]', 'primary_id': '223000', 'version': '0', 'description': 'LACTASE DEFICIENCY, CONGENITAL;;ALACTASIA, CONGENITAL;;DISACCHARIDE INTOLERANCE II', 'dbname': 'MIM_MORBID', 'synonyms': [], 'info_type': 'DEPENDENT', 'info_text': '', 'db_display_name': 'MIM morbid'}

Reactome gene

{'display_id': 'R-HSA-1430728', 'primary_id': 'R-HSA-1430728', 'version': '0', 'description': 'Metabolism', 'dbname': 'Reactome_gene', 'synonyms': [], 'info_type': 'DIRECT', 'info_text': 'Generated via gene', 'db_display_name': 'Reactome gene'}

Reactome gene

{'display_id': 'R-HSA-71387', 'primary_id': 'R-HSA-71387', 'version': '0', 'description': 'Metabolism of carbohydrates', 'dbname': 'Reactome_gene', 'synonyms': [], 'info_type': 'DIRECT', 'info_text': 'Generated via gene', 'db_display_name': 'Reactome gene'}

Reactome gene

{'display_id': 'R-HSA-189085', 'primary_id': 'R-HSA-189085', 'version': '0', 'description': 'Digestion of dietary carbohydrate', 'dbname': 'Reactome_gene', 'synonyms': [], 'info_type': 'DIRECT', 'info_text': 'Generated via gene', 'db_display_name': 'Reactome gene'}

UniGene

{'primary_id': 'Hs.551506', 'dbname': 'UniGene', 'ensembl_identity': 99, 'synonyms': [], 'ensembl_start': 1, 'xref_start': 1, 'xref_end': 6274, 'db_display_name': 'UniGene', 'display_id': 'Hs.551506', 'ensembl_end': 6274, 'version': '0', 'score': 31370, 'cigar_line': '6274M', 'description': 'Lactase', 'xref_identity': 100, 'evalue': None, 'info_text': '', 'info_type': 'SEQUENCE_MATCH'}

UniProtKB Gene Name

{'display_id': 'LCT', 'primary_id': 'P09848', 'version': '0', 'description': '', 'dbname': 'Uniprot_gn', 'synonyms': [], 'info_type': 'DEPENDENT', 'info_text': '', 'db_display_name': 'UniProtKB Gene Name'}

WikiGene

{'display_id': 'LCT', 'primary_id': '3938', 'version': '0', 'description': 'lactase', 'dbname': 'WikiGene', 'synonyms': [], 'info_type': 'DEPENDENT', 'info_text': '', 'db_display_name': 'WikiGene'}

You will find different kinds of databases, such as vertebrate and genome annotation project (VEGA), UniProt (Using the Protein Data Bank), or WikiGene.

You can nd a detailed explanation of all the functionalities available at

http://rest.ensembl.org/

This includes all the interfaces and Python code snippets, among other languages.

If you are interested in paralogues, this information can be retrieved quite trivially from the preceding recipe. On the call to homology/id, just replace the type with paralogues.

SUMMARY

In this chapter, you have learnt to handle genomics datasets. Many more methods exist, which are beyond the scope of this book. Interested reader is recommended to follow the references for better start points.

REFERENCES

1. Bioinformatics with Python by Tiago Antao (Packt Publishing, 2015).

8

PHYLOGENY

Modern science has shown that all species of organisms that live on earth undergo a slow transformation process through the ages. We call this process *evolution.* One of the central problems in biology is to explain the evolutionary history of today's species and, in particular, how species relate to one another in terms of common ancestors. This is usually done by constructing trees, whose leaves represent present-day species and whose interior nodes represent hypothesized ancestors. These kinds of trees are called *phylogenetic trees*. According to this tree human beings and chimpanzees are genetically closer to each other than to the other primates in the tree. This means that they have a common ancestor that is not an ancestor of any of the other primates. The tree also shows a similar situation between gibbons and siamangs.

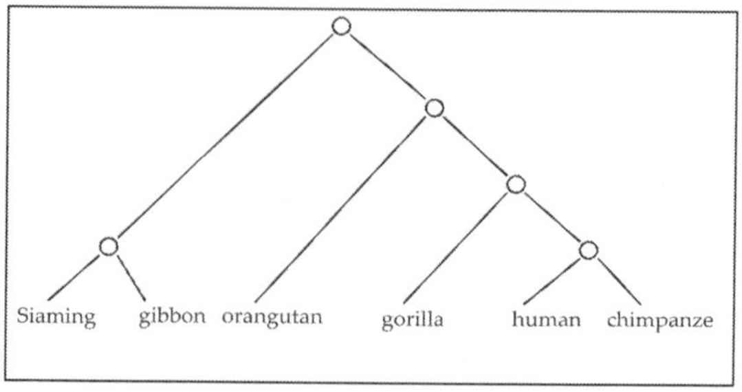

Figure 8.1: Primate phylogenetic tree.

The problem with phylogenetic tree construction is that generally we do not have enough data about distant ancestors of present-day species; and even if we did, we could not be 100% sure that a particular fossil belongs to a species that *really* is the ancestor of two or more current species. Therefore we have to *infer* the evolutionary history of current organisms, and *re-create* their phylogenetic tree, generally using as primary data comparisons between today's species. For this reason the tree shown in Figure 8.1 is not necessarily the true one; it is simply a hypothesis.

Phylogenetic trees, or more simply, phylogenies, have been proposed since the last century for all sorts of groups of organisms. The methods used up to the 1950s were generally based on researchers' experience and intuition. Gradually mathematical formalisms were introduced and these were incorporated into numerical methods. Today there is an immense body of work on the mathematics and algorithmics of phylogeny reconstruction. There are also software packages offering a host of programs to aid scientists in re-creating phylogenies.

It is not our aim in this chapter to survey computational methods for phylogeny reconstruction; a whole book would be required for this. Our goal is rather to study some problems of this area that are amenable to algorithmic solution. The reader should be aware that this is a very narrow and biased treatment, considering the extent and diversity of the literature on this topic.

Scientists are interested in two main aspects of phylogenetic trees. One is its *topology,* that is, how its interior nodes connect to one another and to the leaves. (Some authors use the term *branching pattern* for this.) Another important aspect is the *distance* between pairs of nodes, which can be determined when the tree edges are weighted. This distance is an estimate of the evolutionary distance between the nodes. Depending on many assumptions that we will not discuss, the distance between an interior node (ancestral object) to a leaf (present-day object) can be interpreted as an estimate of the time it took for one to evolve into the other. But the reader should be aware evolutionary distance in general is not equivalent to elapsed time.

Another important feature of a phylogenetic tree is whether it has a root. A tree with a root implies ancestry relationships between interior nodes. However, in many of the problems that we will study there is not enough information in the data to allow us to determine the root; in those cases the reconstructed tree will be unrooted.

SIMULATIONS

Instead of going into the algorithmic complexities of the topics, we adopt a practical approach which involves using simulations to study phylogeny. This should help the reader in understanding the goal and scope of phylogenetic studies. We have borrowed a good deal of code and hands-on part for this chapter from the excellent work at

http://readiab.org/book/latest/

The reader is requested to refer to the above website for more details. Here, we focus on Phylogeny.

```
%pylab inline
import numpy as np
import seaborn as sns
import random
from IPython.core import page
page.page = print
Populating the interactive namespace from numpy and matplotlib
```

First we'll look at the function used to simulate the evolution of a single sequence. This is where most of the important evolutionary modeling of sequence evolution happens.

```
from iab.algorithms import evolve_sequence

%psource evolve_sequence

def evolve_sequence(sequence, substitution_probability, indel_
probability):
  result = []
  sequence_length = len(sequence)
  insertion_choices = list(sequence.nondegenerate_chars)
  result = []
  i = 0
  while i < sequence_length:
    current_char = sequence[i]
    if random.random() < substitution_probability:
        # simulate a substitution event by adding a character other than
the current
      # character to the result at this position
            substituted_base = random.choice([r for r in sequence.
nondegenerate_chars if r != current_char])
      result.append(substituted_base)
      i += 1
    elif random.random() < indel_probability:
      # simulate either an insertion or a deletion event. the length of the
insertion or
        # deletion is determined at random, with shorter lengths being
more probable
      length = int(np.random.triangular(1, 1, 10))
      if np.random.binomial(1, 0.5) == 0:
        # simulate an insertion by adding length random characters from
          # this sequence's alphabet
        result.extend(np.random.choice(insertion_choices, size=length))
```

```
            i += 1
        else:
            # simulate a deletion by not appending any of the next length
            # characters
            i += length
    else:
        # simulate no mutation occurring
        result.append(str(current_char))
        i += 1
return sequence.__class__(''.join(result))
```

Next, take a look at the function that models a single generation of a single sequence. This is where the clonal reproduction (i.e., one parent sequence becoming two child sequences) occurs.

```
from iab.algorithms import evolve_generations
%psource evolve_generations
def evolve_generations(ancestral_sequence, generations, substitution_probability,
                       indel_probability, increased_rate_probability=0.0,
                       fold_rate_increase=5, verbose=False):
    """
    ancestral_sequence : skbio.Sequence object
        The sequence that should be used as the root node in the tree.
    generations : int
        The number of generations to simulate. Must be greater than zero.
    substitution_probability : float
        The probability at which a substitution mutation should occur.
    indel_probability : float
        The probability at which either an insertion or a deletion mutation
        should occur. One of these will be simulated with even probability,
```

```
    and the length will be randomly chosen between 1 and 9, with shorter
     lengths being more probable.
  increased_rate_probability : float, optional
    The probability at which child2 should experience an increased rate
    of substitution mutations. Default is that this never occurs.
  fold_rate_increase : int, optional
    If increased_rate_probability is greater than zero, the fold increase
    that substitutions should now occur at.
  verbose : bool, optional
    If True, print the sequences that are simulated at each generation.
  """

  # initial some values and perform some basic error checking
  assert generations > 0, "Must simulate one or more generations."
  # initialize a list of the previous generations sequences - this gets used
  # in the for loop below. since we'll start with the first generation of
  # children, root_sequence is the previous generation's sequence
  ancestral_sequence.metadata['id'] = '0'
  previous_generation_sequences = [ancestral_sequence]
  # iterate for each requested generation
  for i in range(generations):
    # print the generation number and the current number of sequences
    if verbose:
        print("Generation: %d (Number of parent sequences: %d)" %
(i,2**i))
      print("%s (last common ancestor)" % ancestral_sequence)
      print("")
    # create a list to store the current generation of sequences
    current_generation_sequences = []
    # iterate over the sequences of the previous generation
    for parent_sequence in previous_generation_sequences:
```

```
        # evolve two child sequences - currently the mutation probabilities
are
        # constant, but should update that to change with generations
            r1, r2 = evolve_generation(parent_sequence, substitution_
probability, indel_probability,
                        increased_rate_probability, fold_rate_increase)
        r1.metadata['id'] = '%s.1' % (parent_sequence.metadata['id'])
        r2.metadata['id'] = '%s.2' % (parent_sequence.metadata['id'])
        current_generation_sequences.extend([r1, r2])
        if verbose:
         # if the caller specified verbose output, print the actual sequences
        print("%s (parent id: %s)" % (parent_sequence, parent_sequence.
metadata['id']))
            print("%s (child 1 id: %s)" % (r1, r1.metadata['id']))
            print("%s (child 2 id: %s)" % (r2, r2.metadata['id']))
            print("")
      # current_generation_sequences becomes the next generation's
      # previous_generation_sequences
      previous_generation_sequences = current_generation_sequences
    # upon completion of all generations, return the last generation's
sequences
  return previous_generation_sequences
```

Finally, take a look at our entry point function. This is where we provide the parameters of our simulation, including the starting sequence, the number of generations, and the mutation probabilities. Notice how each of these functions builds on the prior functions.

Now we'll run our simulation. We'll start with a random DNA sequence, and then evolve three generations. Before running this, can you predict how many child sequences we'll end up with after three generations? When we call evolve_generations, we'll pass the parameter verbose=True. This will tell the function to print out some information throughout the process. This will let us inspect our evolutionary process: something that is impossible to do with real sequences.

from iab.algorithms import random_sequence

import skbio

sequence = random_sequence(skbio.DNA, 50)

sequences = evolve_generations(sequence, generations=3, substitution_probability=0.1,

indel_probability=0.05, increased_rate_probability=0.1, verbose=True)

Generation: 0 (Number of parent sequences: 1)

AAGAGACGTAGCTGCTAGCTTTATGTTAATCCATTGC-GGGAGACACGCAT (last common ancestor)

AAGAGACGTAGCTGCTAGCTTTATGTTAATCCATTGC-GGGAGACACGCAT (parent id: 0)

AAGAGACGTAGCTGCTGCTTTACGTTAATCCATTGAGGAGA-ACCACGCAT (child 1 id: 0.1)

AGACGCTGTCAACTTTATGTCAATCCATTGCGGCAGACACG-CAT (child 2 id: 0.2)

Generation: 1 (Number of parent sequences: 2)

AAGAGACGTAGCTGCTAGCTTTATGTTAATCCATTGC-GGGAGACACGCAT (last common ancestor)

AAGAGACGTAGCTGCTGCTTTACGTTAATCCATTGAGGAGA-ACCACGCAT (parent id: 0.1)

AAGAGAAGCTGCTGTTTTAAGACCTTTTATGCCATTCAG-GAGAACCACGCAT (child 1 id: 0.1.1)

AAGAGACGTAGGTCCTTCTACGTTAATCCATTGAGGAGAAC-CACTCAT (child 2 id: 0.1.2)

AGACGCTGTCAACTTTATGTCAATCCATTGCGGCAGACACG-CAT (parent id: 0.2)

AGACGCTGTCAACTGCCAATCCATCGGGTGCGGCAGACT (child 1 id: 0.2.1)

ATACGCTGTCAACTTTATGTCCATCCATTGCGCAGACACGCAT (child 2 id: 0.2.2)

Generation: 2 (Number of parent sequences: 4)

AAGAGACGTAGCTGCTAGCTTTATGTTAATCCATTGC-GGGAGACACGCAT (last common ancestor)

AAGAGAAGCTGCTGTTTTAAGACCTTTTATGCCATTCAG-GAGAACCACGCAT (parent id: 0.1.1)

ATGAGAAGCTGCTGTTTTAAGACCTTTTATGCGAATCAGGT-GAACCACGCAT (child 1 id: 0.1.1.1)

AAGGGAAGCTGCTGGTTAAGACCTTTTATGCCATTCAGGAGA-CCCACACAT (child 2 id: 0.1.1.2)

AAGAGACGTAGGTCCTTCTACGTTAATCCATTGAGGAGAAC-CACTCAT (parent id: 0.1.2)

AAGAGATTTAGGTCCTACGTTTCCTTGGGCACTCAT (child 1 id: 0.1.2.1)

ATGAGACGTAGGTCCTTCTACGTTACTTTCCATTGAGGAACG-GTCGACTAAACTTCAT (child 2 id: 0.1.2.2)

AGACGCTGTCAACTGCCAATCCATCGGGTGCGGCAGACT (parent id: 0.2.1)

GGACGCGGTCAACTGCCTATGCACCGCGTGCTATCAG-CAGACT (child 1 id: 0.2.1.1)

AGACGCTGTCAACCGCCAATCCATCGGGTTCCGCAGACC (child 2 id: 0.2.1.2)

ATACGCTGTCAACTTTATGTCCATCCATTGCGCAGACACGCAT (parent id: 0.2.2)

ATACGCTGTGAAGACATTTATGTACATCCATTGAGCAGGATTG-CACGTGAT (child 1 id: 0.2.2.1)

ATACGCTGTAAACCTTATGTCCATCTATTGCCTAGACACGCAT (child 2 id: 0.2.2.2)

We now have a new variable, sequences, which contains the child sequences from the last generation. Take a minute to look at the ids of the parent and child sequences above, and the ids of a couple of the final generation sequences. These ids are constructed so that each sequence contains the identifiers of its ancestral sequences, and then either 1 or 2. Notice that all sequence identifiers start with 0, the identifier of the last common ancestor (or our starting sequence) of all of the sequences. These identifiers will help us interpret whether the phylogenies that we reconstruct accurately represent the evolutionary relationships between the sequences.

Also, notice that at this point we only have the sequences from the last generation. We no longer have the ancestral sequences (which would correspond to the internal nodes in the tree). This models the real world, were we only have sequences from extant organisms, but not their ancestors.

Take a minute to compare the two sequences below. What types of mutations happened over the course of their evolution?

```
print(len(sequences))

8

sequences[0]

DNA

---------------------------------------------------------

Metadata:

    'id': '0.1.1.1'

Stats:

    length: 52

    has gaps: False

    has degenerates: False
```

```
   has non-degenerates: True
   GC-content: 42.31%
-------------------------------------------------------------
0 ATGAGAAGCT GCTGTTTTAA GACCTTTTAT GCGAATCAGG TGAACCACGC AT
sequences[-1]
DNA
---------------------------------------------------
Metadata:
   'id': '0.2.2.2'
Stats:
   length: 43
   has gaps: False
   has degenerates: False
   has non-degenerates: True
   GC-content: 41.86%
---------------------------------------------------
0 ATACGCTGTA AACCTTATGT CCATCTATTG CCTAGACACG CAT
```

PITFALLS OF SIMULATIONS

While simulations are extremely powerful for comparing algorithms, they can also be misleading. This is because when we model evolution we simplify the evolutionary process. One assumption that our simulation is making is that "bursts" of evolution (i.e., where our substitution rate temporarily increases) are restricted to only a single generation. A child is no more or less like to have an increased rate of substitutions if its parent did. This may or may not be an accurate model. Imagine in the real-world that the environment changed drastically for some descendants (for example, if a geological event created new thermal vents in a lake they inhabited, resulting in an increase in mean water temperature), but not for others. The lineages who experience the environmental change might have an increased rate of substitutions as their genomes adapt to the new environment. The increased substitution rate may persist for multiple generations, or it may not. This is one limitation of our simulation that we know about, but because we don't have a perfect understanding of sequence evolution, there are limitations that we don't know about.

When using simulations, it's important to understand what assumptions the simulation makes so you know what it can tell you about and what it can't tell you about. You'll want to consider that when determining how confident you are in the results of an evaluation based on simulation. What are some other assumptions that are being made by the evolutionary simulation presented here? There are many, so take a minute to list a few.

On the opposite end of the spectrum from simulations for algorithm comparison is comparisons based on real data. The trade-off is that, while there are no assumptions being made in real data with respect to the composition of the sequences, with real data we don't know what the right answer (in our case, the correct phylogeny) is, so it's harder to determine which algorithms are doing better or worse. The take-away message here is that neither approach is perfect, nor often researchers will use a combination of simulated and real data to evaluate algorithms.

VISUALIZE TREES

As we now start computing phylogenetic trees, we're going to need a way to visualize them. We'll use the *ete3* Python package for this, and in the next cell we'll configure the *TreeStyle* which is used to define how the trees we visualize will look. If you'd like to experiment with other views, you can modify the code in this cell according to the *ete3* documentation.

```
import ete3

ts = ete3.TreeStyle()

ts.show_leaf_name =True

ts.scale =250

ts.branch_vertical_margin =15
```

We can apply this TreeStyle to a random tree as follows. Any changes that you make to the TreeStyle above will impact this tree and all following trees in this chapter. Experiment with this - it's fun!

```
t = ete3.Tree()
t.populate(10)
t.render("%%inline", tree_style=ts)
```

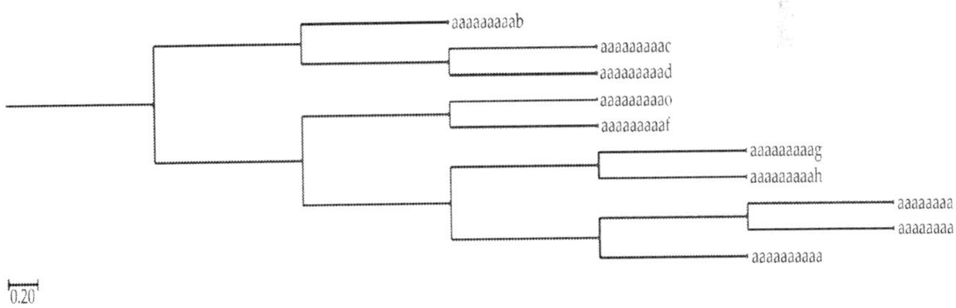

DISTANCES AND DISTANCE MATRICES

Technically speaking, a distance between a pair of objects is a measure of their dissimilarity. There isn't one single definition of the distance between two objects. For example, if your two objects are the cities Flagstaff, Arizona and Boulder, Colorado you might measure the distance between them as length of shortest line that connects them. This would be the Euclidean distance between the two cities. If you're trying to travel from one city to another however, that might not be the most relevant distance measure. Instead, you might be interested in something like their Manhatten distance, which in this would be more similar to a measure of the shortest route that you could travel between the two cities using the Interstate Highway system.

Similarly, different distance metrics will be relevant and not relevant for different types of objects. Clearly the way you measure distances between cities is very different than the way we measure distances between biological sequences. However, the underlying concept of a distance between two objects is the same.

When we compute the distances between some number of objects *n*, we'll commonly represent those values in a distance matrix which contains all of those values. These distance matrices are so common in bioinformatics that scikit-bio defines a DistanceMatrix object that provides a convenient interface for working with these data. We can create one of these objects as follows:

```
from skbio import DistanceMatrix
dm = DistanceMatrix([[0.0, 1.0, 2.0],
                     [1.0, 0.0, 3.0],
                     [2.0, 3.0, 0.0]],
                     ids=['a', 'b', 'c'])
```

We can then access the values in the distance matrix directly, view the distance matrix as a heatmap, and do many other things that facilitate analyzing distances between objects.

```
print(dm['a', 'b'])
print(dm['b', 'c'])
1.0
3.0
_ = dm.plot(cmap='Greens')
```

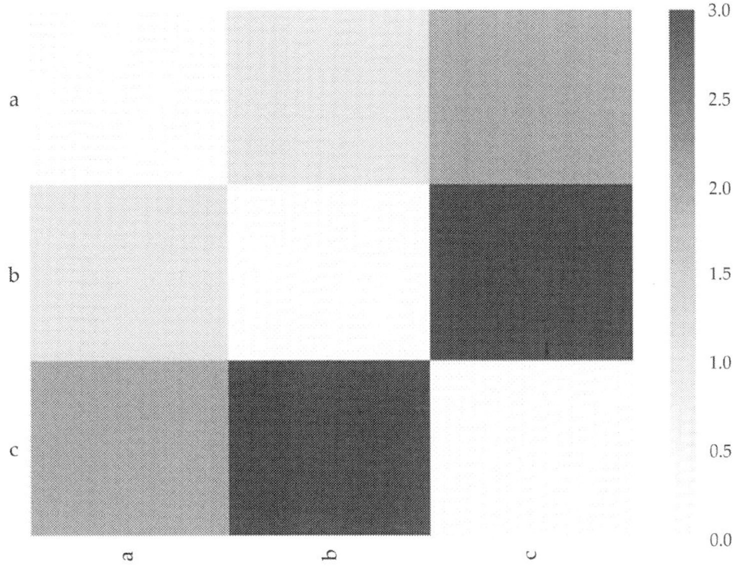

ALIGNMENT-FREE DISTANCES BETWEEN SEQUENCES

We've now looked at several ways of computing distances between sequences, some of which have required that the positions in the sequences are directly comparable to one another (i.e., that our sequences are aligned), and some of which haven't. One alignment-free distance between sequences is the k-mer distance.

We can use the kmer_distance function with scikit-bio as follows to create an skbio.DistanceMatrix object.

```
from iab.algorithms import kmer_distance

kmer_dm = DistanceMatrix.from_iterable(sequences, metric=kmer_distance, key='id')

_ = kmer_dm.plot(cmap='Blues', title='3mer distances between sequences')
```

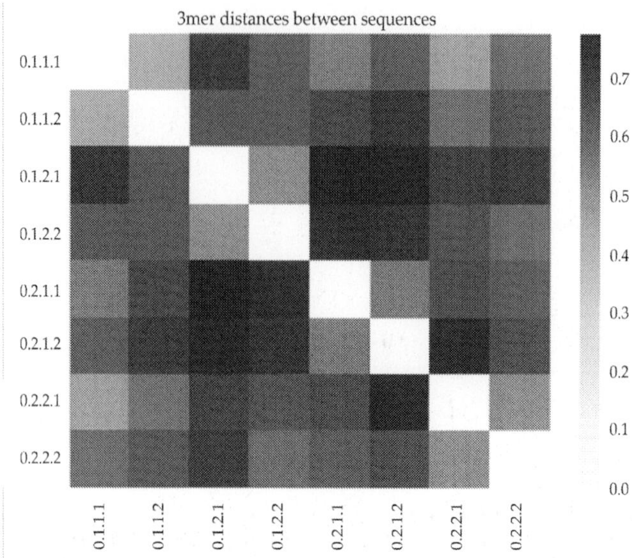

ALIGNMENT-BASED DISTANCES BETWEEN SEQUENCES

One alignment-based distance metric that we've looked at is Hamming distance. This would be considered an alignment-based approach because is does consider the order of the characters in the sequence by comparing a character at a position in one sequence only to the character at the corresponding position in the other sequence. We could compute these distances as follows, after first aligning our sequences.

```
from skbio.alignment import global_pairwise_align_nucleotide
from iab.algorithms import progressive_msa
from functools import partial
indel_probability =0.0
sequences = evolve_generations(sequence, generations=10, substitution_probability=0.03,
                        indel_probability=0.0, increased_rate_probability=0.1, verbose=False)
sequences = random.sample(sequences, 25)
```

```
if indel_probability ==0:
    sequences_aligned = sequences
else:
    gpa = partial(global_pairwise_align_nucleotide, penalize_terminal_gaps=True)
    sequences_aligned = progressive_msa(sequences, pairwise_aligner=gpa)
from skbio.sequence.distance import hamming
hamming_dm = DistanceMatrix.from_iterable(sequences_aligned, metric=hamming, key='id')
_ = hamming_dm.plot(cmap='Oranges', title='Hamming distances between sequences')
```

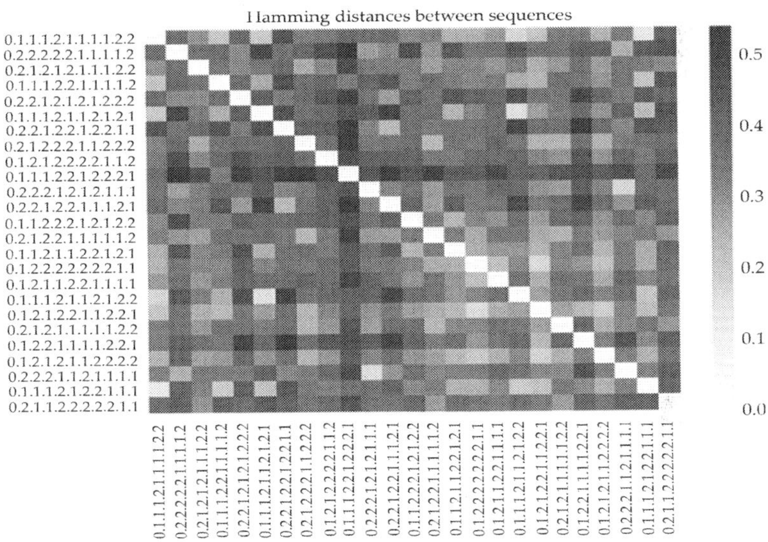

JUKES-CANTOR CORRECTION OF OBSERVED DISTANCES BETWEEN SEQUENCES

The Python implementation of this correction looks like the following. We can apply this to a number of input distance values to understand how it transforms our Hamming distances.

```
def jc_correction(p):
```

```
return (-3/4) * np.log(1- (4*p/3))
distances = np.arange(0, 0.70, 0.05)
jc_corrected_distances = list(map(jc_correction, distances))
ax = sns.pointplot(distances, jc_corrected_distances)
ax.set_xlabel('Hamming distance')
ax.set_ylabel('JC-corrected distance')
ax.set_xlim(0)
ax.set_ylim(0)
ax
<matplotlib.axes._subplots.AxesSubplot at 0x118cc3a20>
```

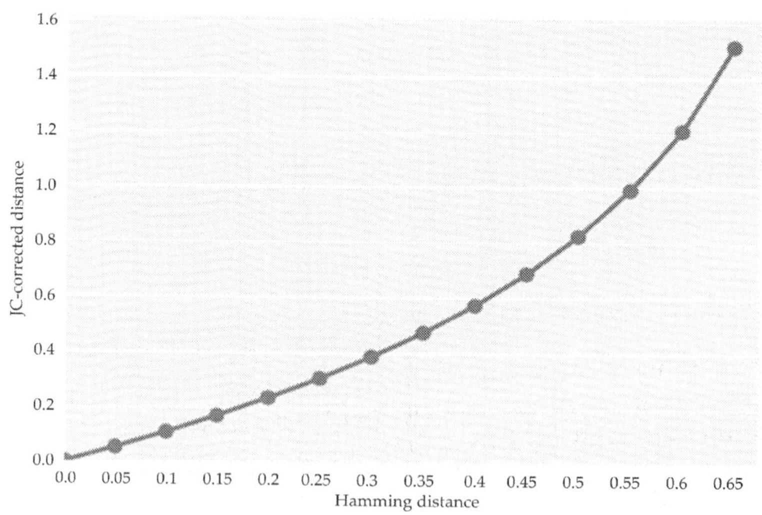

We can then apply this to a full distance matrix as follows (we'll then print the first row of each).

```
def jc_correct_dm(dm):
  result = np.zeros(dm.shape)
for i in range(dm.shape[0]):
for j in range(i):
        result[i,j] = result[j,i] = jc_correction(dm[i,j])
return skbio.DistanceMatrix(result, ids=dm.ids)
```

```
jc_corrected_hamming_dm = jc_correct_dm(hamming_dm)
print(hamming_dm[0])
print(jc_corrected_hamming_dm[0])
[ 0.   0.4  0.3  0.2  0.4  0.18 0.46 0.32 0.34 0.38 0.36 0.38
  0.3  0.32 0.24 0.24 0.28 0.16 0.2  0.34 0.34 0.24 0.34 0.12
  0.42]
[ 0.         0.57160504 0.38311922 0.2326162  0.57160504 0.20582763
  0.71264421 0.417216   0.45293704 0.52992765 0.49044485
0.52992765
  0.38311922 0.417216   0.28924686 0.28924686 0.35050538 0.179963
  0.2326162  0.45293704 0.45293704 0.28924686 0.45293704
0.13076504
  0.61573541]
_ = jc_corrected_hamming_dm.plot(cmap='Greens', title='JC-corrected
Hamming distances between sequences')
```

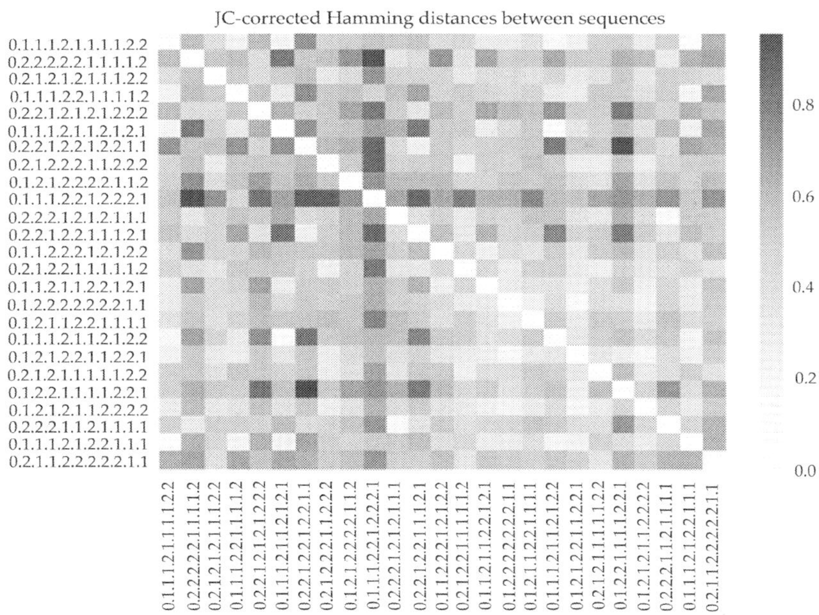

SUMMARY

In this chapter, you have learnt common methods employed in Phylogenetic studies by way of tutorials. The methods are presented as illustrations, as this prepares the reader to delve into the details.

REFERENCES

1. Introduction to Computational Genomics: A Case Studies Approach by Cristianini N & Hahn MW (Cambridge University Press, 2006)
2. Molecular Evolution: A Phylogenetic Approach by Page RDM & Holmes EC (Wiley-Blackwell, 2009).
3. Little Book of R for Bioinformatics by Avril Coghlan, Wellcome Trust Sanger Institute, Cambridge, U.K (2016).
4. Bioinformatics with Python by Tiago Antao (Packt Publishing, 2015).

9

PROTEIN STRUCTURE PREDICTION

We introduced the problem of *predicting* protein structure from sequence in the introductory chapter. If one thinks carefully, most of the problems we address computationally are some form of predictions. Here, we explore one such method which has a formal presentation in terms of dynamic programming. The method is called THREADER. This chapter differs from previous ones, in that it is more theoretical and would invite and or attract the attention of people who can implement a theoretical formulation in the form of an algorithm or a program. It also invites the attention of people from other fields, as predicting a structure from a sequence is a fundamental problem in biophysics and other allied interdisciplinary areas.

DEFINITIONS

We start by introducing some definitions as are described in the initial theoretical formulation of the protein structure and folding from its sequence. This involves using some mathematics. It is enough if you know how functions are written and their description. The prediction of protein tertiary structure from sequence may be expressed symbolically by expressing the folding process as a mathematical function:

C=F(S)

where

$$S=[s_1, s_2,\ldots, s_n],\ C=[\theta_1, \theta_2,\ldots, \theta_{3n-2}],\ s\in\{Ala, Arg,\ldots, Val\}$$

In this case the main chain conformation of the protein chain ***S*** is represented as a *vector* of main chain torsion angles ***C,*** with the chain itself being defined as a vector of elements corresponding to members of the set of 20 standard amino acids. With this simplistic definition we have provided a mathematical footing for describing the polypeptide chain conformation in terms of mathematical functions.

The **folding** process, hence, can be defined as a *function* which takes an amino acid sequence and computes from it a sequence of main chain torsion angles. The polypeptide chain conformation, as a choice, can be represented in torsional space arbitrarily, and the problem can just as readily be expressed in terms of relative orthogonal 3D coordinates. The protein folding problem can be considered as a search for the folding function ***F.*** It is probable, however, that no simple representation of the folding function exists, and that even if the function exists in any form, the only device capable of performing the required biological function is the protein chain itself. Conceptually, the simplest way to arrange for a protein sequence to code for its own native structure is to arrange for the sequence to adopt a structure with a global minimum of the protein chain's free energy. The folding process is therefore transformed into an *energy function minimization* process, where the energy function has input the protein sequence vector ***S,*** and the vector of torsion angles ***C*** . Given a particular sequence ***S,*** the folding process becomes a *search* through the set of all corresponding vectors of torsion angles ***C*** for the minimum of an energy function *E,* where *E* is defined as:

$$E(S,C_{\_native}<E(S,C_{\_(non\text{-}native})})$$

The exact form of this energy function is as yet unknown. We start with a few reasonable assumptions. It is reasonable is to assume that it would incorporate terms pertaining to the types of interactions observed in protein structures, such as hydrogen bonding and van der Waals effects among many others. The conceptual simplicity of this model for protein folding stimulated much research into *ab initio* tertiary structure prediction. Teritiary structure prediction, *ab initio*, requires solutions of two mathematical problems. These are outlined below.

- Find a potential function for which the above inequality *at least* generally holds.
- Construct an **algorithm** capable of finding the *global minimum* of this function.

To date, these problems remain essentially unsolved, though a lot of progress has been made, particularly with the construction of efficient minimization algorithms. It is unlikely that proteins really locate the global minimum of a free energy function in order to fold into their native conformation. The case against proteins searching conformational space for the global minimum of free energy was argued by Levinthal. The *Levinthal paradox***,** as it is now known, can be demonstrated fairly easily. If we consider a protein chain of ***N*** residues, we can estimate the size of its conformational space as roughly 10^N states. This assumes that:

The main chain conformation of a protein may be completely represented by a suitable choice from just 10 main chain torsion angle triplets for each residue.

This of course neglects the additional conformational space provided by the *side chain torsion angles*, but is a reasonable rough estimate. The paradox comes from estimating the *time* required for a protein chain to search its conformational space for the global energy minimum.

Let us take an example of a protein chain of length 100 residues. It is clear that no physically achievable search rate would enable this chain to complete its folding process. Even if the atoms in the chain were able to move at the speed of light, it would take the chain around 10^{82} seconds to search the entire conformational space, which compares rather unfavorably to the estimated age of the Universe (10^{17} seconds).

Clearly proteins do not fold by searching their entire conformational space. There are many ways of explaining Levinthal's paradox.

A highly plausible mechanism for protein folding is that of encoding a *folding pathway* in the protein sequence. Despite the fact that chains of significant length cannot find their global energy minimum, short chain segments (5-7 residues) could quite easily locate their global energy minimum within the average lifetime of a protein, and it is therefore plausible that the location of the native fold is driven by the folding of such *short fragments*.

Levinthal's paradox is only a paradox if the energy surface formed by the free energy function is highly convoluted, with no downhill paths leading to the global minimum. Short fragments can be envisaged to traverse a small downhill segment of the free energy surface during folding, and if these paths were to converge on the global energy minimum then the protein would have located its *native* fold. One subtle point to make about the relationship between the minimization of a protein's free energy and protein folding is that

the native conformation need not correspond to the global minimum of free energy.

One possibility is that the folding pathways initially locate a *local* minimum, but a local minimum which provides stability for the average lifetime of the protein. In this case, the protein in question would always be observed with a free energy higher than the global minimum ***in vivo,*** but would eventually locate its global minimum if isolated and left long enough ***in vitro*** - though the location of the global minimum could take longer time. Thus, a biologically active protein could in fact be in a ***metastable*** state, rather than a stable one. Let us now pay some attention to the folds.

LIMITED FOLDS

Evidence in investigations targeted at protein folding studies point towards an observation that there are a limited number of *naturally occurring* protein folds. If we consider a chain of length 50 residues we might calculate the number of possible main chain conformations as 7^{50} ($\sim 10^{42}$). Clearly, most of these conformations will not be stable folds and many will not be even physically possible. In order to form a compact globular structure, a protein chain necessarily has to form regular secondary structures, and it is this constraint, along with the constraints imposed from a requirement to effectively pack the secondary structures formed, that limit the number of stable conformational states for a protein chain. In addition to the constraints imposed from physical effects on protein stability, there are also evolutionary constraints on the number of occurring folds.

Where do new proteins come from?

The answer is of course from other proteins. In other words the folding patterns we observe today are the result of the evolution of a set of ancestral protein folds. If the number of possible folds is limited, then this fact should be apparent in the presently known protein structures.

Do folds recur in apparently unrelated proteins?

The answer appears to be a definite "yes". Reports of these "fold analogies" are common in the literature, though whether this is due the probability of the fold of a newly solved structure matching an existing one increases due to the increase in the number of known folds (*saturation* effect), or whether this is simply due to an increased awareness of the possibility (and the increased use of structural comparison programs) is a matter of debate and research.

Protein structure prediction is thus presented with a "short-cut" which arises from the observation that the number of possible folds are limited and not uncountable. As already described, it is impractical to attempt tertiary structure prediction by searching a protein's entire conformational space for the minimum energy structure. But, if we know that there could be as few as 1000 possible protein folds, then the intelligent way to search a protein's conformational space would be to simply consider only those regions which correspond to this predefined set. This is analogous to the difference between an exam requiring the writing of an essay and an exam requiring multiple-choice questions to be answered. Clearly a person with no knowledge of the subject at hand has a much greater chance of achieving success with the multiple-choice paper than with the essay paper.

Suppose we had derived a practical potential function for which the native conformational energy was lower than that of any other conformation, and that we had identified ***F*** possible chain folds, then we would have the basis of a useful tertiary structure prediction scheme. In order to predict the conformation of a given protein chain ***S***, the chain would be folded into each of the ***F*** known chain conformations (C_l, ..., C_M), and the energy of each conformation calculated. The predicted chain conformation would be the conformation with the lowest value of the potential function. The term generally applied to schemes of this type is *fold recognition,* where instead of trying to predict the fold of a protein chain *ab initio,* we attempt to recognize the correct chain fold from a list of alternatives.

LIBRARIES OF FOLDS

Certainly, this points to the fact there exists a library of possible folds of protein tertiary structure. A suitable representative library of folds needs to be found and catalogued for a quick reference. These folds can be observed in crystal structures, *NMR* (nuclear magnetic resonance) structures, or even theoretical model structures. If the library is limited to observed structures then the method will evidently be capable of recognizing only previously observed folds. Any new structure prediction, would have to checked against this library of folds for suitable local matches. As has been already discussed, the frequency of occurrence of similar folds between proteins sharing no significant sequence similarity would seem to indicate that creating a library entirely out of *known* folds is perfectly reasonable.

MODELLING PROCESS

The central process in structure-*based fold recognition* is the fitting of a given sequence onto a structural template.

One way of visualizing this process is to imagine the side chains of the object protein being fitted onto the backbone structure of the template protein. This process is of course almost identical to the process of comparative modelling. The standard modelling process consists of three basic steps.

- Initially *at least* **one** suitable *homologous template* structure needs to be found.
- An *optimal alignment* needs to be generated between the sequence of the template structure (the *subject* sequence) and the sequence of unknown structure (the *query* sequence).

- The framework structure of the template is "mapped" onto the query sequence. After several stages of energy minimization, the model is ready for critical evaluation.

Each step in the modelling process has its associated problems, though the first two steps are the most critical overall. If no homologous structure can be found, the process cannot even be started, and even when a homologous structure is available (perhaps selected on the basis of functional similarity), the degree of homology may be so low as to render the alignment of the sequences impossible by normal means ("by eye" or by automatic alignment).

The general assumption is that some residual sequence similarity exists between the template sequence and the sequence under investigation, which is often not the case. The ideal modelling method would not make this assumption, and work with cases where there is no detectable sequence similarity between the object sequence and the source protein. We now turn our attention to pairwise potentials.

STATISTICALLY DERIVED PAIRWISE POTENTIALS

Several groups have used statistically derived **painvise potentials** to identify incorrectly folded proteins. A succesful attempt by Bowies is decribed here.

The first stage of this method involves the prediction of *residue accessibility* from multiple sequence alignments. In essence, alignment positions with high average hydrophobicity and high conservation are predicted to be buried and relatively polar variable positions predicted to be exposed to solvent. The degree of predicted exposure at each position of the aligned sequence family is then encoded as a string. This string is then matched against a library of similarly encoded strings, based, however, not on predicted accessibilities but on *real* accessibilities calculated from structural data.

Despite the obvious computational advantages of using *residue environments*, it is clear that the fold of a protein chain is governed by fairly specific **protein-protein** and **protein-solvent** atomic interactions. A given protein fold is therefore better modelled in terms of a "network" of pairwise interatomic energy terms, with the structural role of any given residue described in terms of its interactions. Classifying such a set of interactions into one environmental class such as "buried alpha-helical" will inevitably result in the loss of useful information, reducing the ***specificity*** of sequence-structure matches evaluated in this way.

The main difficulty in the use of environments alone for recognizing protein folds is that helices look like other helices, and strands like other strands. A sequence that folds into one helix of particular structure, will probably easily fold into any other helix of similar length. A very good example of two topologies which cannot be distinguished after encoding into environmental classes is $(\alpha\beta)_8$ barrel (a "TIM barrel") and a parallel αβ sandwich (a Rossmann fold). In this case both topologies comprise alternating α and β structure, where the strands are mostly inaccessible to solvent. Providing that the αβ sandwich is of sufficient size, or if flanking domain regions provide additional secondary structural elements (the Rossmann domain itself typically has only 6 strands), then the one-dimensional descriptors of the two structures are almost identical.

The factor that limits the scope of the search for a stable threading is **packing**. Whilst the sequence of any isolated helix could substitute for any other, the sequences for a packed pair of helices are much more highly constrained. For a complete protein structure, solvation effects also come into play. In general, for a globular protein the threading of its sequence onto its structure is constrained by:

- *local interactions* (in the example given, the required formation of a helix),
- *long-range pairwise interactions* (helix-helix packing for example) and
- *solvation effects*, which are primarily governed by the periodic accessibilities of exposed helices and strands.

We should like to match a sequence to a structure by considering the plethora of detailed pairwise interactions, rather than averaging them into a crude environmental class. Incoporation of such a procedure into an alignment algorithm like *Needleman-Wunsch* is impractical. Certainly, an approach which involves optimization of individual terms seems to be the solution. We spend some time on evaluating the model bilt using the above approach theoretically.

MODEL EVALUATION FUNCTION

The general approach described here employs a set of *information theoretic potentials* of mean force. These potentials associate event probabilities with statistical free energy. If a certain event is observed with probability *p* (say the occurrence of a valine residue α-carbon and an alanine α-carbon at a separation of 5 Å) we can associate an "energy" with this event using the inverse Boltzmann formula:

$$E= -kT \ln (p)$$

The constant *-kT* may be ignored, in which case the units are no longer those of free energy but of ***information***. For simplicity, we have also ignored the additional term *Z*, known as the Boltzmann sum, a clear explanation of why this is acceptable is given by Sippl. The important point about both *free energy* and *information entropy* formulations of probability is that the resulting values are *additive*.

Consider two independent events with probabilities *p* and *q*, respectively. The probability of both events occurring together is simply *pq*, but multiplication is difficult to implement in pattern matching algorithms. Transforming the combined probability *pq* by taking *logs* provides the following useful result:

$$\ln(pq)=\ln(p)+\ln(q)$$

Therefore the important part of the calculation of potentials of mean force, and the related techniques of information theory is simply converting probabilities to *log-likelihoods*.

The real computational key to this approach to transforming probabilities into energy-like parameters, is really so that we can transform the problem of multiplying probabilities into the problem of adding related terms. In general it is relatively easy to handle additive terms algorithmically, but very hard to handle values which need to be multiplied. If it makes it easier to follow, readers who are happier working with units of information than units of energy can simply erase the *-kT* terms from the equations which follow, and even change the base of the logarithms to *base 2* so that the scores can be expressed in units of "bits".

We are interested in *relative* rather than *absolute* probabilities. Taking the above example, it is of little interest to know how probable it is that a valine α-carbon and an alanine α-carbon are found to be separated by 5 Å. Of much greater interest is the question of how probable this valine-alanine separation is in comparison with other residue pairs. If the probability of any residue pair having an α -carbon separation of s is $f(s)$ and the frequency of occurrence for residue pair ab is $f_{ab}(s)$ then we can write down the potential of mean force as follows:

$$\Delta E_{ab}(s) = -kT \ln \left[\frac{f_{ab}(s)}{f(s)} \right]$$

Sippl divides this potential into a set of potentials relating to different topological levels 1, . . . , k, which is simply the residue pair sequence separation.

For the tripeptide sequence MFP (Met-Phe-Pro), $k = 1$ for residue pairs MF (Met-Phe) and FP (Phe-Pro), with k = 2 for residue pair MP (Met-Pro). In reality, probability density functions $f_k(s)$ and $f_k^{ab}(s)$ are unknown and must be replaced by the relative frequencies observed in the available structural database denoted $g_k(s)$ and $g_k^{ab}(s)$, respectively, where s is typically divided into 20 intervals for sampling. As there are 400 residue pairs (sequence asymmetry is assumed) and only some 15000- 20000 residues in the set of non-homologous protein structures, the observed frequency distributions $g_k^{ab}(s)$ are only weak approximations of the true probability densities and must therefore be corrected to allow for the very small sample size. By considering the observation process as the collection of information quanta, Sippl suggests the following transformation:

$$f_k^{ab}(s) \approx \frac{1}{1 + m\sigma} g_k(s) + \frac{m}{1 + m\sigma} g_k^{ab}(s)$$

where m is the number of pairs ab observed at topological level k and is the weight given to each observation. As m this transformation has the required property that the right and left-hand sides of the equation become equal as $g_k^{ab}(s)$ $f_k^{ab}(s)$.

$$\Delta E_k^{ab} = kT \ln(1 + m\sigma) - kT \ln \left[1 + m_{ab} \frac{g_k^{ab}(s)}{g_k(s)}\right.$$

A possible criticism of the mean force potentials proposed by Sippl is that there exists in the force-field a dependence on protein size. The problem lies in the fact that interactions even as distant as 80 Å are taken into account in the calculation of the potentials, and so consequently, the bulk of data for these large distances is derived from large proteins. The ideal case would be for the potentials to be calculated from proteins of roughly equal size to the protein of interest. Unfortunately, this simple solution is generally impractical. The data set used to generate the mean force potentials is already sparse, even before subdivision into size ranges.

In order to render the mean force potentials less dependent on protein chain length, these long-distance interactions must be replaced by a size independent parameter. The first requirement in replacing these interaction parameters is to determine a suitable dividing line which separates short-distance from long-distance interactions. The next step is then to determine the nature of the information encoded by these interactions. Finally, a suitable size-independent parameter can be sought to replace this information

Consider two protein atoms separated by a distance *d*. Clearly if *d* is large there will be no significant physical interaction between these atoms. Conversely, if *d* is small then we might expect there to be some influence, whether it be a hydrophobic effect, an electrostatic effect or even a covalent interaction. If such an influence exists, then we might also expect there to be some residue preferences for the residues containing these atoms, and consequently we would expect some kind of correlation between the two residue identities. This provides a possible way to determine a cut-off distance for meaningful residue-residue interactions. If the identities of two residues can be considered to be independent variables, then these residues (or more correctly, the residue side chains) will probably not be involved in a significant physical interaction.

To determine the degree of dependency between residues separated by a particular distance, some measure of statistical association is required. The method selected here is based on *statistical entropy*, a common concept in statistical physics and information theory. The entropy of a system with I states, where each state occurs with probability p_i, is defined as:

$$H(x) = -\sum_{i=1}^{I} p_i \ ln(p_i)$$

Consider two experiments x and y with I and J possible outcomes respectively, each of which occurs with a probability p_i $(i= 1,\ldots,I)$, and p_j $(j=1,\ldots,J)$.

The entropy ***H*** of these systems is defmed as:

$$H(x) = -\sum_{i=1}^{I} p_i \; ln(p_i)$$

$$H(x) = -\sum_{j=1}^{J} p_j \; ln(p_j)$$

Entropy in this case is essentially defined as the degree of freedom of choice, or more strictly in this case, the degree of equiprobability. If the outcome probabilities in each experiment are equal then the statistical entropy is maximized, as this represents the maximum freedom of choice. If the probability of one outcome is unity (the others of course being zero) then zero entropy is achieved, corresponding to a lack of choice whatsoever.

If we link both experiments, then we can represent the overall outcomes in the form of a *contingency table*. An example of such a linked pair of experiments is the throwing of a pair of dice, in which case the contingency table would have 6 rows and 6 columns, representing the 6 possible outcomes for each die.

The entropy of the combined experiment is:

$$H(x, y) = -\sum_{i,j} p_{ij} \; ln(p_{ij})$$

To determine the statistical association between experiments ***x*** and ***y***, the entropy of *y given x* and *x given y* may be derived. If a knowledge of the outcome of experiment x allows a wholly accurate prediction of the outcome of experiment y, then the entropy of *y given x* must be zero. Conversely, if a knowledge of experiment x is found to be of no benefit whatsoever in the prediction of y, then the conditional entropy in this case is maximized.

The entropy of *y given x* is as follows:

$$H(y|x) = \sum_{i,j} p_{ij} \ln\left(\frac{p_{ij}}{p_i}\right)$$

and the entropy of *x given y* is:

$$H(y|x) = \sum_{i,j} p_{ij} \ln\left(\frac{p_{ij}}{p_j}\right)$$

Finally a suitable symmetric measure of interdependence (known as the ***uncertainty)*** between ***x*** and ***y*** is defined thus:

$$U(x,y) \equiv 2\frac{H(y) + H(x) - H(x,y)}{H(x) + H(y)}$$

An uncertainty between *x* and *y* of zero indicates that the two experimental variables are totally independent, whereas an uncertainty of one indicates that the two variables are totally dependent. One would hope that in the case of the two dice experiment previously described, that *U(x, y)* would be found to be close to zero for a large number of trials, though gluing the dice together would be a sure way of forcing *U(x,y)* to unity.

Using the uncertainty measure, it is now possible to evaluate residue correlations in protein structures. In this case, the two experimental variables are the identities of two residues separated by a given distance in a particular structure. Effects are observed across short sequential separations, where steric and covalent effects predominate. Most importantly, both the short- and long-range interactions become undetectable when averaged over distances greater than around 12 Å (though it must first be realized that for the very short separations, 1 and 2, it is impossible for the separation to exceed 10 Å). It must be stressed that this does not necessarily imply that the physical forces themselves do not act beyond this distance, only that the effects do not manifest themselves in the selection of amino acid residues.

CALCULATION OF POTENTIALS

For the short-range potentials, minimum and maximum painvise distances were determined for each type of atomic pairing at each topological level from a small set of very highly resolved crystal structures. These distance ranges were subdivided into 20 intervals. For the medium and long-range potentials, interactions were sampled over the range 0-10 Å with a fixed sampling interval of 2 Å.

This potential simply measures the frequency with which each amino acid species is found with a certain degree of solvation, approximated by the residue solvent accessible surface area. The solvation potential for amino acid residue ***a*** is defined as follows:

$$\Delta E^{a}_{solv}(r) = -kT\ln\left[\frac{f^{a}(r)}{f(r)}\right]$$

where ***r*** is the **%** residue accessibility (relative to residue accessibility in GGXGG extended pentapeptide). Residue accessibilities can be calculated using the DSSP program.

OPTIMAL THREADING SEARCH

Given an efficient means for the evaluation of a hypothetical sequence threading relationship, the problem of finding the optimal threading must be considered. For a protein sequence of length ***L*** and a template structure of which ***M*** residues are in regular secondary structures, the total number of possible threadings is given by:

$$\binom{L}{M} \equiv \frac{L!}{(L-M)!M!}$$

One way to reduce the scale of the problem is to restrict insertions and deletions (indels) to the loop regions. By excluding indels from secondary structural elements, the problem reduces to a search for the optimal set of loop lengths for a protein. Under these conditions threading a sequence on a structure may be visualized as the sliding of beads on an abacus wire, where the beads are the secondary structures, and the exposed wire the remaining loop regions.

Unfortunately, even with this extreme restriction on the threading patterns, the search space again becomes unreasonably large for proteins of even average size. The exact number of threadings with secondary structure indels disallowed is a complex function of both sequence length and the number and lengths of constituent secondary structural elements. As a rule of thumb, however, it is found that for N secondary structural elements, with loops of average length (say between 2 and 12 residues), the number of threadings is $O(10^N)$. For typical proteins comprising 1&20 secondary structures, it is clearly not possible to locate the optimal threading by an exhaustive search in a reasonable period of time.

COMBINATORIAL OPTIMIZATION

Some methods are described below:

Exhaustive search

As demonstrated earlier, for small proteins of **<6** secondary structural elements, and disallowing secondary structure indels, it is practical to simply search through all possible threadings in order to locate the threading of lowest energy. Unfortunately, the evaluation function is tailored towards average sized globular proteins with hydrophobic cores, whereas the small proteins tend to be less globular and typically lack a hydrophobic core. This is called Exhasutive search.

Monte carlo methods

Monte Carlo methods have been often exploited for conformation calculations on proteins. Simulated annealing has been exploited in the alignment of protein structures, and in the optimization of side chain packing in protein structures. Simulated annealing is a simple random search process. In this instance, random threadings are generated and evaluated using the evaluation function described earlier. Where a proposed threading has a lower energy than the current threading, the proposed threading is accepted. In the case where a proposed threading has a higher energy than the current, it is accepted with probability p , where

$$p = exp\left(-\frac{\Delta E}{kT}\right)$$

and where E is the difference between the current and the proposed threading energy and T is the current annealing "temperature". After a predefined number of accepted changes, the temperature is slightly reduced. This whole procedure is repeated until no further reduction in threading energy is achieved, at which point the system is said to be frozen. The schedule of cooling is critical to the success of simulated annealing.

We also have dynamic programming methodologies and we defer the discussion as the complexity is beyond the scope of the present book.

RESIDUE SELECTION FOR SEQUENCE-STRUCTURE ALIGNMENTS

The residue selection stage of optimal sequence threading involves the summation of local interaction potential terms $E^{a}{}_{k}b$ over all residue pairs in overlapping windows of length ***L,*** where ***L*** is a small constant odd-number over the range, say, 5-31. Similarly the solvation terms are are summed for each of the ***L*** residues. The window is clipped appropriately if it spans either the N- or C-terminus of either the sequence or the structure, for example the window length for the first and last residue in either cases would be ***:(L*** + 1). To equalize the contribution of the pairwise terms and the solvation terms, the average energy is calculated and summed for both, giving a total energy for the sequence-structure fragment of:

$$S_{mn} = E\left(fragmnet\right) = \frac{2\sum_{i=1}^{L-1}\sum_{j=i+1}^{L} E_{pair}^{ij}}{L\left(L-1\right)} + \frac{\sum_{k=1}^{L} E_{solv}^{k}}{L}$$

Energies are calculated for every sequence fragment threaded onto every structure fragment, and the results stored in a selection matrix S. Using the residue selection step, the number of initial pairs *(m,n)* for which low-level paths need to be calculated is reduced up to 100-fold.

SOFTWARE

A program, called THREADER, which implements the threading approach to protein fold recognition described here has been made widely available to the academic community free of charge, and can be downloaded over the Internet from the URL:

http://bioinf.cs.ucl.ac.uk/?id=747

REFERENCES

1. Jones, D.T., Taylor, W.R. & Thornton, J.M. (1992) A new approach to protein fold recognition. *Nature*. 358, 86-89.
2. Jones, D.T., Miller, R.T. & Thornton, J.M. (1995) Successful protein fold recognition by optimal sequence threading validated by rigorous blind testing. *Proteins*. 23, 387-397.
3. Levinthal, C. (1968) Are there pathways for protein folding ? *Chim. Phys*. 65, 4445.
4. Needleman, S.B. and Wunsch, C.D. (1970) A general method applicable to the search for similarities in the amino acid sequence of two proteins. *J. Mol. Biol.* 48, 443453.
5. Sippl, M.J. (1990) Calculation of conformational ensembles from potentials of mean force. An approach to the knowledge-based prediction of local structures in globular proteins. *J. Mol. Biol.* 213, 859-883.
6. Casari, G., Sippl, M.J (1992). Structure-derived hydrophobic potential. Hydrophobic potential derived from X-ray structures of globular proteins is able to identify native folds. *J. Mol. Biol.* 224, 725-732.
7. Bowie, J.U., Clarke, N.D., Pabo, C.O. and Sauer, R.T (1990). Identification of protein folds: matching hydrophobicity patterns of sequence sets with solvent accessibility patterns of known structures. *Proteins* 7, 257-264.
8. Bowie, J.U., Luthy, R. and Eisenberg, D (1991). Assessment of protein models with three-dimensional profiles. *Science* 253, 164-170.

APPENDIXS

Appendix 1

We introduce FASTQ in this appendix. FASTQ is a common format for storing sequencing reads. FASTA and FASTQ are rather similar, but FASTQ is almost always used for storing *sequencing reads* (with associated quality values), whereas FASTA is used for storing all kinds of DNA, RNA or protein sequencines (without associated quality values). Before delving into the format, I should mention that there are great tools and libraries for parsing and manipulating FASTQ, e.g. [FASTX], and [BioPython]'s [SeqIO] module. If your needs are relatively simple, you might try using these tools and libraries and skip reading this document.

[FASTA]: http://en.wikipedia.org/wiki/FASTA_format

[FASTQ]: http://en.wikipedia.org/wiki/FASTQ_format

[BioPython]: http://biopython.org/wiki/Main_Page

[SeqIO]: http://biopython.org/wiki/SeqIO

[FASTX]: http://hannonlab.cshl.edu/fastx_toolkit/

Basic format

Here's a single sequencing read in FASTQ format:

@ERR294379.100739024 HS24_09441:8:2203:17450:94030#42/1

AGGGAGTCCACAGCACAGTCCAGACTCCCAC-CAGT TCTGACGAAATGATGAGAGCTCAGAAGTAACAGTTGCTT-TCAGTCCCATAAAAACAGTCCTACAA

+

BDDEEF?FGFFFHGFFHHGHGGHCH@ GHHH GFAHEGFEHGEFGHCCGGGFEG FGFFDFFHBGDGFHGEFGHFGH GFGFFFEHGGFGGDGHGFEEHFFHGE

#

It's spread across four lines. The four lines are:

#

1. "`@`" followed by a read name

2. Nucleotide sequence

3. "`+`", possibly followed by some info, but ignored by virtually all tools

4. Quality sequence (explained below)

Here is a very simple Python function for parsing file of FASTQ records:

In[1]:

```
def parse_fastq(fh):
    """ Parse reads from a FASTQ filehandle. For each read, we
      return a name, nucleotide-string, quality-string triple. """
    reads = []
    while True:
      first_line = fh.readline()
      if len(first_line) == 0:
        break  # end of file
```

```
        name = first_line[1:].rstrip()
        seq = fh.readline().rstrip()
        fh.readline()  # ignore line starting with +
        qual = fh.readline().rstrip()
        reads.append((name, seq, qual))
    return reads
```

```
fastq_string = '''@ERR294379.100739024 HS24_09441:8:2203:17450:94030#42/1
AGGGAGTCCACAGCACAGTCCAGACTCCCACCAGTTCTGACGAAATGATG
+
BDDEEF?FGFFFHGFFHHGHGGHCH@GHHHGFAHEGFEHGEFGHCCGGGF
@ERR294379.136275489 HS24_09441:8:2311:1917:99340#42/1
CTTAAGTATTTTGAAAGTTAACATAAGTTATTCTCAGAGAGACTGCTTTT
+
@@AHFF?EEDEAF?FEEGEFD?GGFEFGECGE?9H?EEABFAG9@CDGGF
@ERR294379.97291341 HS24_09441:8:2201:10397:52549#42/1
GGCTGCCATCAGTGAGCAAGTAAGAATTTGCAGAAATTTATTAGCACACT
+
CDAF<FFDEHEFDDFEEFDGDFCHD=GHG<GEDHDGJFHEFFGEFEE@GH'''
from StringIO import StringIO
parse_fastq(StringIO(fastq_string))
```

The nucleotide string can sometimes contain the character "`N`". `N` essentially means "no confidence." The sequencer knows there's a nucleotide there but doesn't know whether it's an A, C, G or T.

Read names often contain information about:

1. The scientific study for which the read was sequenced. E.g. the string `ERR294379` (an [SRA accession number](http://www.ebi.ac.uk/ena/about/sra_format)) in the read names correspond to [this study](http://www.ncbi.nlm.nih.gov/sra/?term=ERR294379).

2. The sequencing instrument, and the exact *part* of the sequencing instrument, where the DNA was sequenced. See the [FASTQ format](http://en.wikipedia.org/wiki/FASTQ_format#Illumina_sequence_identifiers) wikipedia article for specifics on how the Illumina software encodes this information.

3. Whether the read is part of a *paired-end read* and, if so, which end it is. Paired-end reads will be discussed further below. The `/1` you see at the end of the read names above indicate the read is the first end from a paired-end read.

Quality values

Quality values are probabilities. Each nucleotide in each sequencing read has an associated quality value. A nucleotide's quality value encodes the probability that the nucleotide was *incorrectly called* by the sequencing instrument and its software. If the nucleotide is `A`, the corresponding quality value encodes the probability that the nucleotide at that position is actually *not* an `A`.

Quality values encoded in two senses: first, the relevant probabilities are rescaled using the **Phread** scale, which is a negative log scale. In other words if p is the probability that the nucleotide was incorrectly called, we encode this as Q where

$Q = -10 * \log_{10}(p)$.

For example, if Q = 30, then p = 0.001, which means 1-in-1000 chance that the nucleotide is wrong. If Q = 20, then $p = 0.01$, a 1-in-100 chance. If Q = 10, then $p = 0.1$, a 1-in-10 chance.

Second, scaled quality values are *rounded* to the nearest integer and encoded using [ASCII printable characters](http://en.wikipedia.org/wiki/ASCII#ASCII_printable_characters). For example, using the Phred33 encoding (which is by far the most common), a *Q* of 30 is encoded as the ASCII character with code 33 + 30 = 63, which is "`?`". A *Q* of 20 is encoded as the ASCII character with code 33 + 20 = 53, which is "`5`". And so on.

```
def phred33_to_q(qual):
  """ Turn Phred+33 ASCII-encoded quality into Phred-scaled integer """
  return ord(qual)-33


def q_to_phred33(Q):
  """ Turn Phred-scaled integer into Phred+33 ASCII-encoded quality """
  return chr(Q + 33)


def q_to_p(Q):
  """ Turn Phred-scaled integer into error probability """
  return 10.0 ** (-0.1 * Q)


def p_to_q(p):
  """ Turn error probability into Phred-scaled integer """
  import math
  return int(round(-10.0 * math.log10(p)))
```

To convert an entire string Phred33-encoded quality values into the corresponding *Q* or *p* values, I can do the following:

```
name, seq, qual = parse_fastq(StringIO(fastq_string))[0]
q_string = list(map(phred33_to_q, qual))
p_string = list(map(q_to_p, q_string))
print(q_string)
print(p_string)
```

You might wonder how the sequencer and its software can *know* the probability that a nucleotide is incorrected called. It can't; this number is just an **estimate**. To describe exactly how it's estimated is beyond the scope of this notebook; if you're interested, search for academic papers with "base calling" in the title.

Paired-end reads

Sequencing reads can come in *pairs*. Basically instead of reporting a single snippet of nucleotides from the genome, the sequencer might report a *pair* of snippets that appear *close to each other* in the genome. To accomplish this, the sequencer sequences *both ends* of a longer *fragment* of DNA. Here is simple Python code that mimicks how the sequencer obtains one paired-end read:

Let's just make a random genome of length 1K

import random

random.seed(637485)

genome = ''.join([random.choice('ACGT') for _ in range(1000)])

genome

The sequencer draws a fragment from the genome of length, say, 250

offset = random.randint(0, len(genome) - 250)

fragment = genome[offset:offset+250]

fragment

Then it reads sequences from either end of the fragment

end1, end2 = fragment[:75], fragment[-75:]

end1, end2

And because of how the whole biochemical process works, the second end is always from the opposite strand from the first.

import string

function for reverse-complementing

_revcomp_trans = string.maketrans("ACGTacgt", "TGCAtgca")

def reverse_complement(s):

    return s[::-1].translate(_revcomp_trans)

```
end2 = reverse_complement(end2)
end1, end2
```

FASTQ can be used to store paired-end reads. Say we have 1000 paired-end reads. We should store them in a *pair* of FASTQ files. The first FASTQ file (say, `reads_1.fq`) would contain all of the first ends and the second FASTQ file (say, `reads_2.fq`) would contain all of the second ends. In both files, the ends would appear in corresponding order. That is, the first entry in `reads_1.fq` is paired with the first entry in `reads_2.fq` and so on.

Here is a Python function that parses a pair of files containing paired-end reads.

```
def parse_paired_fastq(fh1, fh2):
  """ Parse paired-end reads from a pair of FASTQ filehandles. For each pair, we return a name, the nucleotide string for the first end, the quality string for the first end, the nucleotide string for the second end, and the quality string for the second end. """
    reads = []
    while True:
      first_line_1, first_line_2 = fh1.readline(), fh2.readline()
      if len(first_line_1) == 0:
        break  # end of file
      name_1, name_2 = first_line_1[1:].rstrip(), first_line_2[1:].rstrip()
      seq_1, seq_2 = fh1.readline().rstrip(), fh2.readline().rstrip()
      fh1.readline()  # ignore line starting with +
      fh2.readline()  # ignore line starting with +
      qual_1, qual_2 = fh1.readline().rstrip(), fh2.readline().rstrip()
      reads.append(((name_1, seq_1, qual_1), (name_2, seq_2, qual_2)))
    return reads

fastq_string1 = '''@509.6.64.20524.149722/1
```

AGCTCTGGTGACCCATGGGCAGCTGCTAGGGAGCCTTCTCTC-CACCCTGA

```
+
H H H H H H H G H H H H H H H H H H H H H H H H H HHHHHHHHHHHHHHHIIHHIHFHHF
@509.4.62.19231.2763/1
G T T G A T A A G C A A G C A T C T C A T TTTGTGCATATACCTGGTCTTTCGTATTC
+
H H H H H H H H H H H H H H E H H H H H H H H HHHHHHHHHHHHHHHDHHHHHHGHGHH'''
fastq_string2 = '''@509.6.64.20524.149722/2
T A A G T C A G G A T A C T T T C C C A T A T CCCAGCCCTGCTCCNTCTTTAAATAAT
+
HHHHHHHHHHH HHHHHHHH@ HHFHHHEFHHHHHHFF#FFFFFFHHHHH
@509.4.62.19231.2763/2
C T C T G C T G G T A T G G T T G ACGCCGGATTTGAGAATCAANAAGAGCTTACTA
+
HHHHHHHHHHHHHHHHHHHEH EHHHFHGHHHHHHHH>@#@=44465HHHHH'''
from StringIO import StringIO
parse_paired_fastq(StringIO(fastq_string1), StringIO(fastq_string2))
```

Other comments

In all the examples above, the reads in the FASTQ file are all the same length. This is not necessarily the case though it is usually true for datasets generated by sequencing-by-synthesis instruments. FASTQ files can contain reads of various lengths. FASTQ files often have extension `.fastq` or `.fq`.

Other resources

* [Wikipedia page for FASTQ format](http://en.wikipedia.org/wiki/Fastq_format)

* [BioPython], which has [its own ways of parsing FASTA](http://biopython.org/wiki/SeqIO)

* [FASTX] toolkit

* [seqtk]

* [FastQC]

[BioPython]: http://biopython.org/wiki/Main_Page

[SeqIO]: http://biopython.org/wiki/SeqIO

[SAMtools]: http://samtools.sourceforge.net/

[FASTX]: http://hannonlab.cshl.edu/fastx_toolkit/

[FASTQC]: http://www.bioinformatics.babraham.ac.uk/projects/fastqc/

[seqtk]: https://github.com/lh3/seqtk

APPENDIX 2

Let us have a look at the SAM format in this section. The **SAM Format** is a text format for storing sequence data in a series of tab delimited ASCII columns. Most often it is generated as a human readable version of its sister BAM format, which stores the same data in a compressed, indexed, binary form. Currently, most SAM format data is output from aligners that read FASTQ files and assign the sequences to a position with respect to a known reference genome. In the future, SAM will also be used to archive unaligned sequence data generated directly from sequencing machines. We offer a sample code for illustrative purpose, but it is advised to use the popular samtools suite to perform your actual analysis. Here's a string representing a three-line SAM file. I'm temporarily ignoring the fact that SAM files usually have several header lines at the beginning.

```
samStr = '"\

r1      0       gi|9626243|ref|NC_001416.1| 18401  42      122M  *
0       0       TGAATGCGAACTCCGGGACGCTCAGTAATGT-
GACGATAGCTGAAAACTGTACGATAAACNGTACGCTGAGGGCAGAAAAAATC-
GTCGGGGACATTNTAAAGGCGGCGAGCGCGGCTTTTCCG
+"@6<:27(F&5)9)"B:%B+A-%5A?2$HCB0B+0=D<7E/<.03#!.
F77@6B==?C"7>;))%;,3-$.A06+<-1/@@?,26">=?*@'0;$:;??G+:#+(
A?9+10!8!?()?7C>     AS:i:-5XN:i:0 XM:i:3XO:i:0 XG:i:0 N M : i : 3
MD:Z:59G13G21G26YT:Z:UU

r2      0       gi|9626243|ref|NC_001416.1| 8886    42      2 7 5 M
*       0       0       NTTNTGATGCGGGCTTGTGGAGTTCAGC-
CGATCTGACTTATGTCATTACCTATGAAATGTGAGGACGCTAT-
GCCTGTACCAAATCCTACAATGCCGGTGAAAGGTGCCGGAT-
CACCCTGTGGGTTTATAAGGGGATCGGTGACCCCTACGC-
GAATCCGCTTTCAGACGTTGACTGGTCGCGTCTGGCAAAAGT-
TAAAGACCTGACGCCCGGCGAACTGACCGCTGAGNCCTAT-
GACGACAGCTATCTCGATGATGAAGATGCAGACTGGACT-
GC      (#!!'+!$""%+(+)'%)%!+!(&++)''"#"#&#"!'!("%''"("+&%$
%*%%#$%#%#!)*'(#")(($&$'&%+&#%*)*#*%*')(%+!%%*"$%"
#+)$&&+)&)*+!"*)!*!("&&"*#+"&"'(%)*("'!$*!!%$&&&$!!&&"
(*"$&"#&!$%'%"#)$#+%*+)!&*)+(""#!)!%*#"*)*')&")($+*%%)!
*)!('(%"'+%"$##"#+(('!*(($*'!"*(""+)&%#&$+('**$$&+*&!#%)')
'(+(!%+         AS:i:-14        XN:i:0 XM:i:8XO:i:0 XG:i:0 N M : i : 8
MD:Z:0A0C0G0A108C23G9T81T46         YT:Z:UU

r3      16      gi|9626243|ref|NC_001416.1| 11599  4               2
338M  *       0       0       GGGCGCGTTACTGGGATGATC-
GTGAAAAGGCCCGTCTTGCGCTTGAAGCCGCCCGAAA-
GAAGGCTGAGCAGCAGACTCAAGAGGAGAAAAATGCG-
CAGCAGCGGAGCGATACCGAAGCGTCACGGCTGAAATATAC-
CGAAGAGGCGCAGAAGGCTNACGAACGGCTGCAGACGC-
CGCTGCAGAAATATACCGCCCGTCAGGAAGAACTGAN-
CAAGGCACNGAAAGACGGGAAAATCCTGCAGGCGGATTA-
CAACACGCTGATGGCGGCGGCGAAAAAGGATTATGAAGC-
GACGCTGTAAAAGCCGAAACAGTCCAGCGTGAAGGTGTCTGC-
GGGCGAT    7F$%6=$:9B@/F'>=?!D?@0(:A*)7/>9C>6#1<6:C(.
```

```
CC;#.;>;2'$4D:?&B!>689?(0(G7+0=@37F)GG=>?958.D2E04C<E,
*AD%G0.%$+A:'H;?8<72:88?E6((CF)6DF#.)=>B>D-="C'B080E'
5BH"77':"@70#4%A5=6.2/1>;9"&-H6)=$/0;5E:<8G!@::1?2DC7C
*;@*#.1C0.D>H/20,!"C-#,6@%<+<D(AG-).?&#0.00'@)/F8?B!&"
170,)>:?<A7#1(A@0E#&A.*DC.E")AH"+.,5,2>5"2?:G,F"D0B8D-
6$65D<D!A/38860.*4;4B<*31?6     AS:i:-22     XN:i:0 X M : i : 8
XO:i:0 XG:i:0 NM:i:8MD:Z:80C4C16A52T23G30A8T76A41
YT:Z:UU"'
```

I'll read this string in line-by-line as though it were a file. I'll (lightly) parse the alignment records.

```
import string
from StringIO import StringIO  # reading from string rather than file
for ln in StringIO(samStr):
   qname, flag, rname, pos, mapq, cigar, rnext, \
   pnext, tlen, seq, qual, extras = string.split(ln, '\t', 11)
   print qname, len(seq) # print read name, length of read sequence
def cigarToList(cigar):
   '" Parse CIGAR string into a list of CIGAR operations.  For more
      info on CIGAR operations, see SAM spec:
      http://samtools.sourceforge.net/SAMv1.pdf '"
   ret, i = [], 0
   op_map = {'M':0, # match or mismatch
           '=':0, # match
           'X':0, # mismatch
           'I':1, # insertion in read w/r/t reference
           'D':2, # deletion in read w/r/t reference
           'N':3, # long gap due e.g. to splice junction
           'S':4, # soft clipping due e.g. to local alignment
           'H':5, # hard clipping
           'P':6} # padding
   Seems like = and X together are strictly more expressive than M. Why
```

not just have = and X and get rid of M? Space efficiency, mainly. The titans discuss: http://www.biostars.org/p/17043/

```
    while i < len(cigar):
        run = 0
        while i < len(cigar) and cigar[i].isdigit():
            # parse one more digit of run length
            run *= 10
            run += int(cigar[i])
            i += 1
        assert i < len(cigar)
        # parse cigar operation
        op = cigar[i]
        i += 1
        assert op in op_map
        # append to result
        ret.append([op_map[op], run])
    return ret
cigarToList('10=1X10=')
def mdzToList(md):
    ''' Parse MD:Z string into a list of operations, where 0=match,
        1=read gap, 2=mismatch. '''
    i = 0;
    ret = [] # list of (op, run, str) tuples
    while i < len(md):
        if md[i].isdigit(): # stretch of matches
            run = 0
            while i < len(md) and md[i].isdigit():
                run *= 10
                run += int(md[i])
```

```
            i += 1 # skip over digit
        if run > 0:
            ret.append([0, run, “”])
    elif md[i].isalpha(): # stretch of mismatches
        mmstr = “”
        while i < len(md) and md[i].isalpha():
            mmstr += md[i]
            i += 1
        assert len(mmstr) > 0
        ret.append([1, len(mmstr), mmstr])
    elif md[i] == “^”: # read gap
        i += 1 # skip over ^
        refstr = “”
        while i < len(md) and md[i].isalpha():
            refstr += md[i]
            i += 1 # skip over inserted character
        assert len(refstr) > 0
        ret.append([2, len(refstr), refstr])
    else:
            raise RuntimeError(‘Unexpected character in MD:Z: “%d”’ %
md[i])
    return ret
```

Each element in the list returned by this call is itself a list w/ 3 elements. Element 1 is the MD:Z operation (0=match, 1=mismatch, 2=deletion). Element 2 is the length and element 3 is the relevant sequence of nucleotides from the reference.

```
mdzToList(‘10A5^AC6’)


def cigarMdzToStacked(seq, cgp, mdp_orig):
  ‘’’ Takes parsed CIGAR and parsed MD:Z, generates a stacked
```

alignment: a pair of strings with gap characters inserted (possibly) and where characters at at the same offsets are opposite each other in the alignment. Only knows how to handle CIGAR ops M=XDINSH right now. '''

```
mdp = mdp_orig[:]
rds, rfs = [], []
mdo, rdoff = 0, 0
for c in cgp:
    op, run = c
    skipping = (op == 4 or op == 5)
    assert skipping or mdo < len(mdp)
    if op == 0:  CIGAR op M, = or X
        # Look for block matches and mismatches in MD:Z string
        mdrun = 0
        runleft = run
        while runleft > 0 and mdo < len(mdp):
            op_m, run_m, st_m = mdp[mdo]
            run_comb = min(runleft, run_m)
            runleft -= run_comb
            assert op_m == 0 or op_m == 1
            rds.append(seq[rdoff:rdoff + run_comb])
            if op_m == 0:  match from MD:Z string
                rfs.append(seq[rdoff:rdoff + run_comb])
            else:  mismatch from MD:Z string
                assert len(st_m) == run_comb
                rfs.append(st_m)
            mdrun += run_comb
            rdoff += run_comb
            Stretch of matches in MD:Z could span M and I CIGAR ops
            if run_comb < run_m:
```

```
            assert op_m == 0
            mdp[mdo][1] -= run_comb
        else:
            mdo += 1
    elif op == 1:  CIGAR op I
        rds.append(seq[rdoff:rdoff + run])
        rfs.append("-" * run)
        rdoff += run
    elif op == 2:  D
        op_m, run_m, st_m = mdp[mdo]
        assert op_m == 2
        assert run == run_m
        assert len(st_m) == run
        mdo += 1
        rds.append("-" * run)
        rfs.append(st_m)
    elif op == 3:  N
        rds.append("-" * run)
        rfs.append("-" * run)
    elif op == 4:  S
        rds.append(seq[rdoff:rdoff + run].lower())
        rfs.append(' ' * run)
        rdoff += run
    elif op == 5:  H
        rds.append('!' * run)
        rfs.append(' ' * run)
    elif op == 6:  P
        raise RuntimeError("Don't know how to handle P in CIGAR")
    else:
        raise RuntimeError('Unexpected CIGAR op: %d' % op)
```

```
    assert mdo == len(mdp)
    return ''.join(rds), ''.join(rfs)
```

Following example includes gaps and mismatches

cigarMdzToStacked('GGACGCTCAGTAGTGACGATAGCTGAAAACCCTGTACGATAAACC', cigarToList('12M2D17M2I14M'), mdzToList('12^AT30G0'))

Following example also includes soft clipping (CIGAR: S), SAM spec: Soft clipping: "clipped sequences present in SEQ". We print them in lowercase to emphasize their clippedness

cigarMdzToStacked('GGACGCTCAGTAGTGACGATAGCTGAAAACCCTGTACGAGAAGCC', cigarToList('12M2D17M2I8M6S'), mdzToList('12^AT25'))

Following example also includes hard clipping (CIGAR: H), SAM spec: Hard clipping: "clipped sequences NOT present in SEQ"

cigarMdzToStacked('GGACGCTCAGTAGTGACGATAGCTGAAAACCCTGTACGAGAAGCC', cigarToList('12M2D17M2I8M6S3H'), mdzToList('12^AT25'))

Note: don't see hard clipping in practice much. Following example also includes skipping (CIGAR: N), as seen in TopHat alignments

cigarMdzToStacked('GGACGCTCAGTAGTGACGATAGCTGAAAACCCTGTACGAGAAGCC',

cigarToList('12M2D10M10N7M2I8M6S3H'),

mdzToList('12^AT25'))

```
def cigarize(rds, rfs):
    off = 0
    oplist = []
    lastc, cnt = '', 0
    for i in xrange(0, len(rds)):
        c = None
```

```
    if rfs[i] == ' ':
      c = 'S'
    elif rds[i] == '-' and rfs[i] == '-':
      c = 'N'
    elif rds[i] == '-':
      c = 'D'
    elif rfs[i] == '-':
      c = 'I'
    elif rds[i] != rfs[i]:
      c = 'X'
    else:
      c = '='
    if c == lastc:
      cnt += 1
    else:
      if len(lastc) > 0:
        oplist.append((lastc, cnt))
      lastc, cnt = c, 1
  if len(lastc) > 0:
    oplist.append((lastc, cnt))
  return ''.join(map(lambda x: str(x[1]) + x[0], oplist))
x, y = cigarMdzToStacked('ACGTACGT', cigarToList('8M'), mdzToList('4G3'))
cigarize(x, y)
```

```
x, y = cigarMdzToStacked('GGACGCTCAGTAGTGACGATAGCTGAAAACCCTGTACGAGAAGCC',
            cigarToList('12M2D10M10N7M2I8M6S3H'),
            mdzToList('12^AT25'))
cigarize(x, y)
```

You can also install the **samtools** and perform jobs efficiently.

INDEX